THE CHAT-GPT GOLDMINE

STRIKE IT RICH WITH A.I.

S.P. BLOOM

VERSIFY AI

PART I

UNBOXING A.I.

1

HARNESS THE FUTURE

In the rapidly evolving landscape of business and technology, entrepreneurs are constantly seeking innovative ways to launch new startups, grow their ventures, stay competitive, and remain agile in a dynamic market. That hasn't always been an easy task.

But the emergence of Artificial Intelligence (AI) has transformed the way entrepreneurs can approach business, creating a paradigm shift that enables them to make more informed decisions, reach new heights of efficiency, and unlock unprecedented opportunities for growth.

What does the term "Artificial Intelligence" mean to you? Does it evoke fears of heartless robots taking over the world and enslaving humanity? I'll admit that I once shared those fears, which were perhaps exacerbated by my appetite for dystopian science fiction. But after my own recent experiences with and education about AI, I've come to see it in a different way.

Forget about the word "Artificial," which essentially just means that the technology is not biological, having been manufactured my human beings. Instead, I ask you to focus on the word "Intelligence," but not in the sense of a consciousness that is separate from yours; see it as an *enhancement* of your intelligence. I wrote this book because I believe that AI, in its current form, and for the foreseeable future, is a priceless tool that is ready to serve as a tireless co-producer as you fulfill your creative ideas and entrepreneurial dreams.

In other words, if you can conceive it, AI can not only help you bring it to fruition, but it can also improve upon your original vision. AI has the power to turn your dreams into reality.

Explore the revolutionary world of A.I. and unleash its profit potential

If you're an entrepreneur, you're probably always looking for ways to make your business more efficient, innovative, and competitive. Or, if you're thinking about starting a business or passive income stream, there's no doubt you want to do so with the best possible chances of success. Enter CHAT-GPT, the advanced AI language model which, although still in its youth, is growing and evolving at a rapid but measured pace, and it already has the potential to revolutionize your career, life, and bottom line.

This book is an essential guide to harnessing the power of AI, with a particular focus on the groundbreaking capabilities of Chat-GPT. As the owner of several e-commerce stores and brands, I can personally attest to the fact that using Chat-GPT for just a few short months has already boosted my website traffic and business income to levels that would have normally taken me several years. And, as a filmmaker, writer, and artist, I've discovered that AI is not a threat to my craft but instead offers myriad ways to help me turn my creative visions into reality.

"Everything we love about civilization is a product of intelligence, so amplifying our human intelligence with artificial intelligence has the potential of helping civilization flourish like never before – as long as we manage to keep the technology beneficial."

- Max Tegmark, Physicist and AI Researcher

This book reveals all the lucrative ways in which Chat-GPT and other AI programs have improved my businesses and crafts, plus an array of other AI secrets that I've learned along the way. Whether you wish to create a lucrative side-hustle, launch a profitable startup, revolutionize an existing business, enhance your creativity, or supercharge your life and career, Chat-GPT can propel it towards unparalleled success.

Throughout the book, we will delve into practical examples, use cases, and step-by-step guides that will empower you to harness the full potential of Chat-GPT, whether you're looking to start a new business or seeking ways to improve your existing one. From content creation and copywriting to customer support, market research, social media management, and beyond, Chat-GPT has the potential to revolutionize the way you run your business and foster a culture of innovation and adaptability.

As you embark on this journey, you'll discover how Chat-GPT can not only save you time and resources but also unlock new revenue streams and opportunities for growth. By embracing AI and Chat-GPT, you'll be equipped to navigate the ever-changing business landscape, and ultimately, transform your entrepreneurial vision into a thriving, profitable reality. Welcome to the future of entrepreneurship and creativity—powered by AI and Chat-GPT.

Supercharged Business

If you're an entrepreneur like me, the challenges you face are multifaceted: from ideation to execution, customer engagement to product development, and marketing to logistics. With AI and Chat-GPT at your disposal, however, you can overcome these challenges, streamline processes, and unleash the full potential of your business.

Section one of the book explores the various aspects of AI, specifically Chat-GPT, and illustrates how you can integrate this cutting-edge technology into your business strategies and life, enabling you to stay ahead of the curve and create something profitable and sustainable.

A Guide for Entrepreneurs

Whether you're running or starting an online or real-world business, working for another company, growing your audience as a social media influencer, churning out creative content as a designer or writer, or still deciding on your goals, CHAT-GPT offers a myriad of applications to help you achieve success. Below are just a few of the many ways in which CHAT-GPT can help you achieve your goals and transform your business, services, brand, or life. You'll find more details about these in the later chapters, plus a range of other lucrative ways to harness the power of CHAT-GPT.

Idea Generation and Business Planning

When starting and growing a business, one of the most challenging tasks is generating ideas and creating a solid business plan. CHAT-GPT can help you brainstorm potential niches, products, or services, and even provide suggestions for unique selling points. Moreover, it can assist you in outlining your business plan by providing templates or guiding you through the process of writing key sections like the executive summary, market analysis, and marketing strategy.

Content Creation and Copywriting

Content is king in the digital age, and crafting high-quality, engaging content can be time-consuming. With CHAT-GPT, you can quickly generate written content, such as blog posts, articles, and social media captions, tailored to your target audience. Additionally, the AI model can help you craft persuasive copy for advertisements, product descriptions, and email campaigns, making your marketing efforts more effective.

Social Media Management

Social media is a crucial channel for promoting your business and engaging with customers. CHAT-GPT can assist you in managing your social media accounts by suggesting content ideas, drafting posts, and even responding to comments and messages. By automating these tasks, you'll save time and ensure consistent communication with your audience.

Customer Support

Providing exceptional customer support can be a game-changer for your business. CHAT-GPT can be integrated into your customer support channels, such as live chat, email, and social media, to handle routine inquiries and resolve issues more efficiently. This allows your team to focus on more complex or sensitive customer interactions, ensuring a high level of customer satisfaction.

Market Research

Understanding your market, competition, and target audience is vital for any business. CHAT-GPT can help you gather valuable insights by analyzing customer reviews, social media conversations, and industry reports. These insights can then be used to refine your marketing strategy, product offerings, and overall business approach.

Product Development

Innovation is key to staying ahead of the competition, and CHAT-GPT can support you in this area by generating ideas for new products, services, or features. It can even help you draft product specifications, user manuals, and other technical documentation, streamlining your product development process.

Personal Branding and Influencer Marketing

If you're a social media influencer or are looking to grow your personal brand, CHAT-GPT can provide invaluable assistance. From generating content ideas to crafting engaging captions and stories, CHAT-GPT can help you consistently create content that resonates with your audience. Additionally, it can help you identify potential brand collaborations, sponsorship opportunities, and even generate pitches to reach out to potential partners.

Search Engine Optimization (SEO)

Effective SEO can increase your website's visibility on search engines. Chat-GPT can help optimize website content, meta tags, and other SEO elements, driving organic traffic and increasing online exposure.

Networking

Building connections is vital for entrepreneurial success. Chat-GPT can help craft personalized outreach messages and networking emails, fostering valuable relationships with potential partners, investors, and customers.

Time Management

Effective time management is key to entrepreneurial productivity, and to live a more fulfilling, productive life. Chat-GPT can help prioritize tasks, set goals, and create schedules, ensuring that time is used efficiently and effectively.

Scaling and Growth

Expanding your business requires strategic planning and execution. Chat-GPT can help identify growth opportunities, create expansion plans, and optimize processes, ensuring sustainable and profitable growth.

Personal and Professional Development

Continual learning and growth are essential for success. Chat-GPT can help create personalized learning plans, recommend resources, and provide insights on various topics, fostering personal and professional development.

CHAT-GPT is a powerful tool that entrepreneurs can harness to start, grow, market, and promote their businesses across various industries. By leveraging its capabilities, you can streamline your business processes, make data-driven decisions, and save valuable time and resources. With the help of this book, now is your opportunity to supercharge your business with CHAT-GPT.

2

HISTORY OF ARTIFICIAL INTELLIGENCE

Delve into the world of AI, from its humble beginnings to its modern-day marvels

Artificial intelligence (AI) has become an integral part of our daily lives, shaping the way we communicate, work, and interact with technology. Its pervasive impact has led to a surge of interest in understanding the underlying principles, history, and current state of AI. Let's journey through the development of AI to gain insights into its historical milestones and get an overview of its present-day applications.

The concept of artificial intelligence has captivated human imagination for centuries, with early roots in mythology and literature. However, the formal inception of AI as a scientific discipline took place in the mid-20th century. Key milestones in the history of AI include:

1950s: The Birth of AI

In 1950, British mathematician and computer scientist Alan Turing proposed the Turing Test, a method for determining whether a machine could exhibit intelligent behavior indistinguishable from that of a human. This marked the beginning of serious exploration into AI.

1956: The Dartmouth Conference

The Dartmouth Conference, held in 1956, is considered the birthplace of AI as a distinct field of study. Researchers from diverse disciplines gathered to discuss the potential of developing machines that could simulate human intelligence.

1960s-1970s: Early AI Research and Development

During this period, AI researchers focused on creating rule-based systems and symbolic manipulation techniques, giving rise to early AI programs like SHRDLU and ELIZA.

1980s: Expert Systems and Connectionism

The 1980s saw the emergence of expert systems, which relied on knowledge-based approaches to solve complex problems in specific domains. Simultaneously, the concept of connectionism, inspired by neural networks, gained traction, laying the groundwork for modern deep learning techniques.

1990s-2000s: Machine Learning and the Internet

Machine learning became the dominant approach in AI research during this period. The rise of the internet provided access to vast amounts of data, fueling the development of data-driven AI algorithms and techniques.

2010s-Present: Deep Learning and AI Renaissance

The integration of deep learning techniques with powerful computing resources has led to an AI renaissance, enabling breakthroughs in areas like computer vision, natural language processing, and speech recognition.

AI in the Present Day

Modern AI systems can be broadly classified into two categories: Narrow AI and Artificial General Intelligence (AGI).

Narrow AI

Narrow AI refers to AI systems designed to perform specific tasks without possessing human-like cognitive abilities. Examples of narrow AI include:

- Voice assistants like Siri and Alexa, which use natural language processing and speech recognition to understand and respond to user commands.
- Autonomous vehicles that rely on computer vision, sensor data, and machine learning algorithms to navigate traffic and make driving decisions.
- Recommendation systems employed by platforms like Netflix or Amazon, which analyze user behavior and preferences to offer personalized suggestions.

Artificial General Intelligence (AGI)

Artificial General Intelligence, or AGI, represents the ultimate goal of AI research: to create machines capable of performing any intellectual task that a human can do. AGI remains a theoretical concept, as no existing AI system has achieved this level of intelligence and versatility.

The journey of artificial intelligence, from its inception as a nascent scientific field to its current status as a transformative force, is a testament to the relentless pursuit of knowledge and innovation. As AI continues to evolve and reshape our world, it is essential to understand its historical context and present.

3

THE MYSTERIES AND FUTURE OF AI REVEALED

Get acquainted with the world of AI-powered language models and their potential impact on diverse domains

Chat-GPT is an advanced AI language model developed by OpenAI, designed to understand and generate human-like text. Its sophisticated capabilities have garnered significant attention, with applications ranging from content creation to customer support. However, it's crucial to understand both the strengths and limitations of Chat-GPT to harness its full potential. Let's explore the basics of Chat-GPT, highlight its best use cases, discuss its limitations, and examine the future development of AI language models.

Understanding Chat-GPT

Built on the GPT architecture, Chat-GPT is a powerful tool that leverages natural language processing (NLP) to understand and generate human-like text. It has been trained on a vast corpus of text data, enabling it to comprehend context, respond to prompts, and create coherent content. As a result, Chat-GPT has a wide range of applications in various domains.

Strengths: Where Chat-GPT Excels

1. **Content:** Chat-GPT can generate high-quality content for blogs, articles, social media, and more, saving time and resources for businesses and individuals alike.
2. **Ideas:** From brainstorming business ideas to exploring innovative concepts, Chat-GPT can provide valuable insights and inspiration to fuel creativity.
3. **Customers:** Businesses can use Chat-GPT to develop AI-powered chatbots and virtual assistants, providing efficient and personalized customer service.
4. **Languages:** Chat-GPT can translate text between multiple languages, making it a useful tool for communication and content localization.
5. **Data:** Leveraging its natural language processing capabilities, Chat-GPT can analyze large volumes of text data, such as customer reviews or social media conversations, to provide insights and trends.

Resources: How Chat-GPT is a Game Changer

1. **Access to knowledge:** ChatGPT has access to an enormous knowledge base, making it an invaluable resource for people seeking information on various subjects.
2. **Time-saving:** Busy entrepreneurs and professionals do not always have the time to search for answers to questions. ChatGPT can provide quick, clear, and accurate responses to questions, saving entrepreneurs time which they can apply to other aspects of their busy lives and careers.

3. **Easy to use:** ChatGPT is very simple to use and can be accessed through various platforms such as websites, mobile apps, and messaging platforms like Facebook Messenger and WhatsApp.
4. **Cost-effective:** Hiring a consultant or expert in a specific field is often expensive, but ChatGPT is a cost-effective solution for entrepreneurs seeking information and advice.
5. **Customized responses:** ChatGPT can provide customized responses based on the specific needs of the user. These include information on market trends, business strategies, marketing tactics, and more.
6. **24/7 availability:** ChatGPT is always available, making it convenient for anyone to get the information they need quickly and within seconds.
7. **Up-to-date information:** Although there are limitations in terms of current events, ChatGPT is continuously updated with new information.
8. **Multilingual:** ChatGPT can communicate in multiple languages, making it accessible to people from diverse backgrounds.
9. **Confidentiality:** Entrepreneurs can ask ChatGPT sensitive questions without worrying about confidentiality issues.
10. **Insightful discussions:** ChatGPT can engage in insightful discussions, which helps people gain new perspectives and ideas on various topics, make informed decisions, save time and money, and gain a competitive edge.

Limitations: Understanding Chat-GPT's Constraints

As a developing technology, Chat-GPT is not without its limitations, although these continue to diminish as it adapts and learns. The following are some of the limitations that a user may encounter with Chat-GPT.

1. **Accuracy:** While Chat-GPT generates human-like text, it may produce inaccurate or nonsensical information. Users should verify the content for correctness before relying on it.
2. **Bias:** Chat-GPT may inadvertently generate biased content, as it has been trained on data that may contain inherent biases. Users should be aware of potential bias and take steps to mitigate it.
3. **Lack of Common Sense:** Chat-GPT can occasionally generate responses that lack common sense, showcasing a limitation in its understanding of real-world knowledge.
4. **Verbose Responses:** Chat-GPT may generate overly verbose or repetitive content, requiring users to edit and refine the output for optimal readability.

The Future of Chat-GPT

As AI language models continue to evolve, we can expect several developments in the future:

1. **Improved Accuracy:** With ongoing research and development, future iterations of Chat-GPT are likely to produce more accurate and reliable content.
2. **Reduced Bias:** Efforts to reduce inherent biases in training data will lead to AI models generating more balanced and unbiased content.
3. **Enhanced Contextual Understanding:** Future models will be better at understanding context and maintaining coherence throughout longer conversations or content pieces.
4. **Domain-Specific Expertise:** We may see the development of AI models tailored to specific industries or domains, providing more accurate and relevant information for specialized use cases.

Chat-GPT is a powerful AI language model that offers numerous applications across diverse domains. By understanding its strengths and limitations, users can harness its capabilities effectively while mitigating potential drawbacks.

As research and development in the field of AI continues, we can expect even more sophisticated and versatile language models that will revolutionize the way we communicate, create, and analyze information.

4

MAKING MONEY: STEP BY STEP

Using Chat-GPT to help you make money requires leveraging its potential in various aspects of your business or investment strategy. Here's a step-by-step guide:

1. **Access Chat-GPT:** Sign up for an API key with OpenAI (https://www.openai.com/) to gain access to the Chat-GPT API. Choose a suitable pricing plan and familiarize yourself with the API documentation (https://platform.openai.com/docs/api-reference/introduction).
2. **Identify areas where Chat-GPT can be of assistance:** Examine your current business or investment strategy and identify areas where Chat-GPT could be beneficial. These areas may include content creation, market research, customer support, social media management, or product development.
3. **Develop specific prompts:** Create clear and concise prompts that address your identified areas of need. For example:
4. *"Generate a list of profitable niche business ideas in the online space."*

5. *"Write engaging email copy to promote my online course."*
6. *"Suggest ways to optimize my website for search engines."*
7. **Send prompts to the Chat-GPT API:** Using the API key, connect to the Chat-GPT API and send the prompts you created. You might need programming knowledge or assistance from a developer for this step. Structure your queries effectively, providing context and being specific about the information you need.
8. **Analyze and apply the responses:** Carefully evaluate the responses generated by Chat-GPT, and apply the most relevant and valuable insights to your business or investment strategy.
9. **Monitor progress and adjust your approach:** Regularly track the performance of the implemented strategies and measure their impact on your revenue. Adjust your approach as needed based on the data and insights you gather.
10. **Diversify income streams:** Use Chat-GPT to explore new income-generating opportunities, such as affiliate marketing, online courses, or investment ideas. Continually expand and diversify your income streams to reduce risk and increase revenue potential.
11. **Stay informed:** Keep up-to-date with market trends, technological advancements, and industry news. Use Chat-GPT to stay informed about changes in your niche and adapt your strategies accordingly.
12. **Optimize and scale:** Continuously refine your strategies and processes to improve efficiency and maximize revenue. Use Chat-GPT to identify areas for optimization and explore ways to scale your business or investments effectively.

Remember, while Chat-GPT can provide valuable insights and assistance, it's crucial to combine its capabilities with your own knowledge, research, and decision-making to achieve financial success.

5

PROMPTS FOR PROFITS

When I speak in front of large audiences at conventions and events, my favorite part of every talk is the end. No, not because it's over—I've actually come to enjoy public speaking—but because that's when I accept questions from the audience. I've realized that I give the most valuable information and tell the best stories during these Q&A sessions, often having been reminded of something I meant to tell the audience but had forgotten until a question reminded me. I can also see that the shift from a one-man presentation to a *conversation* enlivens the audience and puts them in maximum learning mode.

In many ways, this is the beauty of Chat-GPT. As an advanced language model, Chat-GPT is ready to have a conversation with you. Your question will most often prompt it to not only provide the information you're seeking, but also a plethora of pertinent information, ideas, instructions, and even inspiration. And, just like in a human-to-human conversation, Chat-GPT will readily answer your follow-up questions with the previous conversation "in mind," so to speak.

In that vein, the Questions and Answers below offer valuable insight about A.I. and Chat-GPT, whether you're a 5-year-old, a technophobe, or technology junkie. Let's start with the 5-year-olds (given their earlier bedtimes).

Chat-GPT Prompts explained to a 5-year-old

Channel your inner child and learn the basics of how to give Chat-GPT a prompt. After all, who's better at asking questions than a 5-year-old?

Q: What is a prompt?

A: A prompt is like a question or idea you give to Chat-GPT. It helps Chat-GPT know what you want it to do, like answering a question or helping you with something.

Q: How do I give Chat-GPT a prompt?

A: You type your question or idea into the box, then Chat-GPT will read it and try to give you a helpful answer or idea back.

Q: How do I make my prompt better?

A: Make your prompt clear and easy to understand. If you need a specific answer or idea, give some details. This helps Chat-GPT know what you want better.

Q: Can I ask Chat-GPT to do anything?

A: Chat-GPT is smart, but it can't do everything. It can help with ideas, answers, and writing, but it can't do things like cook dinner or find your lost toy.

Q: What if Chat-GPT doesn't understand my prompt?

A: If Chat-GPT doesn't understand, try making your prompt clearer or ask your question in a different way. Sometimes, you might need to try a few times before Chat-GPT gives you the answer or idea you want.

Q: How long should my prompt be?

A: Your prompt can be short or long, but it's better to keep it simple. Just make sure you give enough information so Chat-GPT knows what you want.

Q: Can Chat-GPT help me with my homework?

A: Chat-GPT can try to help with some homework, like writing a story or answering a question. But remember, it's important to learn and do your own work too.

Q: Is it okay to ask Chat-GPT silly questions?

A: Yes! Chat-GPT can be fun too. You can ask it to tell you jokes or make up a funny story. Just remember that it's a computer and might not always understand what's funny to you.

Q: Can I use Chat-GPT all day long?

A: You can use Chat-GPT when you need help, but remember to take breaks, play outside, and spend time with your friends and family too. It's important to have a good balance.

Chat-GPT Prompts Explained to a Technophobe

The word "technophobe" is defined as someone who is afraid of technology, although in this case I'm using the word in a broader sense to also include people who have limited experience with the latest technology—which, in this day and age, covers most of us!

To put it another way, this list of Questions and Answers is a resource for anyone who is just diving into the waters of A.I. and Chat-GPT for the first time—and not just for those who still prefer flip-phones over smartphones.

Q: What is a prompt?

A: A prompt is a question or statement you give to Chat-GPT, guiding it on what you want it to generate. Think of it as giving instructions to help the AI understand your needs.

Q: How do I create a good prompt?

A: Be clear and specific with your instructions. For example, instead of asking "Write about dogs," you could ask "Write a short paragraph about the benefits of adopting a rescue dog."

Q: Can I provide more context in my prompt?

A: Yes, providing context helps the AI understand your request better. You can include a brief introduction or context to help guide the response.

Q: How long can my prompt be?

A: Your prompt can be as short or as long as you need, but keep in mind that longer prompts might make the AI's response shorter due to character limitations. Make your prompt concise yet informative.

Q: How do I ask Chat-GPT for multiple ideas?

A: To get multiple ideas, you can use words like "list" or "bullet points." For example: "List 5 creative marketing strategies for a bakery."

Q: Can I ask Chat-GPT to think step by step or debate pros and cons?

A: Yes, you can ask the AI to think step by step or debate pros and cons. For example: "Outline the pros and cons of using social media for marketing a small business."

Q: What if the AI doesn't give me the answer I want?

A: If you're not satisfied with the response, you can try rephrasing your prompt, adding more context, or asking for more details. Experiment with different approaches to get the desired output.

Q: Can I set a specific format or structure for the response?

A: Yes, you can specify the format or structure you want. For example: "Write a 3-paragraph article about the benefits of electric cars, with each paragraph focusing on a different benefit."

Q: Is there any limit to the number of prompts I can use with Chat-GPT?

A: While there might be limits depending on your subscription plan, you can generally use as many prompts as you need to get the information or content you want.

Q: How do I make sure my prompts are safe and appropriate?

A: Always ensure your prompts follow the platform's guidelines and respect others' privacy. Avoid using sensitive information, offensive language, or asking for inappropriate content.

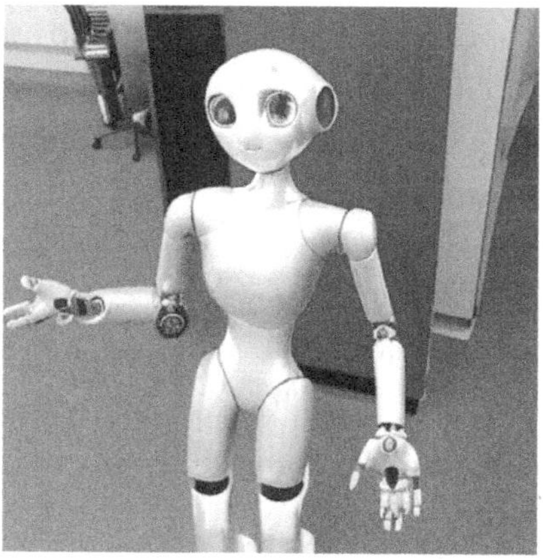

Chat-GPT Prompts Explained To A Tech Junkie

Last but not least, here is a list of questions and answers for people who are already well-versed in technology, including advanced users of Chat-GPT who wish to use prompts for maximum success. If that doesn't yet describe you, it's still worth reading—and definitely worth coming back to this section soon when you are an advanced user.

Q: What's the best way to format prompts for more complex responses from Chat-GPT?

A: Use a clear and concise question, and consider providing context or specifying the desired format of the answer. For example: "Generate a Python script to scrape the top 10 search results for a keyword from Google search."

Q: Can I use variables or tokens within my prompts to create more dynamic conversations with Chat-GPT?

A: Yes, you can use variables or tokens in your prompts to make the conversation more dynamic. Consider using placeholders in your prompts and replace them with the desired values before sending the request to Chat-GPT.

Q: How can I get more accurate and detailed responses from Chat-GPT for technical or niche topics?

A: Make your prompts more specific, and provide context or background information to help guide the model's response. For example: "Explain the differences between GPU and CPU rendering in the context of 3D graphics and game development."

Q: What's the best way to use the temperature and max tokens settings to fine-tune Chat-GPT's output?

A: Lower temperature values (e.g., 0.5) will make the output more focused and deterministic, while higher values (e.g., 1.0) will make it more creative and random. Adjust the max tokens setting to control the length of the response.

Q: How can I use Chat-GPT to generate code snippets or perform code analysis for different programming languages?

A: Craft a prompt that specifies the programming language and the desired action. For example: "Generate a JavaScript function that calculates the factorial of a given number using recursion."

Q: Can Chat-GPT be integrated with other APIs or services to create more sophisticated applications?

A: Yes, you can combine Chat-GPT with other APIs or services to create powerful applications. For example, you could integrate it with a translation API to build a multilingual chatbot.

Q: How can I chain prompts together to create a more interactive conversation with Chat-GPT?

A: Keep track of the conversation history and include it in your subsequent prompts. This will help maintain context and improve the continuity of the conversation. For example, concatenate previous messages with new prompts using a consistent format like "User: {message} \n AI: {response} \n User: {new_prompt}."

Q: Can I use Chat-GPT to generate custom training data for machine learning models or reinforcement learning algorithms?

A: Yes, you can create prompts to generate synthetic training data, but always validate the quality and accuracy of the generated data before using it in your models or algorithms.

Whether a beginner or advanced user, one of the biggest keys to finding success with Chat-GPT is to experiment with different prompt structures and settings. By doing so, you'll soon begin to fine-tune the output to fit your specific needs and likewise uncover profitable surprises and ideas along the way.

6

GET STARTED: STEP BY STEP

Step 1: Open the Internet

First, open your preferred internet browser, such as Google Chrome, Safari, or Firefox, on your computer or device.

Step 2: Visit OpenAI's website

Type "www.openai.com" into the browser's address bar and hit enter. This will take you to OpenAI's website, the company that created Chat-GPT.

Step 3: Navigate to Chat-GPT page

Once you're on the OpenAI website, look for information about Chat-GPT. You might find it in the menu under "Products" or on the homepage. Click the link to navigate to the Chat-GPT page.

Step 4: Create an account or sign in

To start using Chat-GPT, you will need an OpenAI account. If you don't have one, click on "Sign Up" and follow the instructions to create an account. If you already have an account, click on "Sign In" and enter your login information.

Step 5: Choose a plan

Once logged in, you will be prompted to choose a subscription plan. There are usually different options available, such as free or paid plans. Each plan will have different features, such as response length, access to advanced tools, or priority support. Carefully read the plan descriptions and choose the one that best suits your needs.

Step 6: Learn about the interface

After selecting a plan, you will be directed to the Chat-GPT interface. Take some time to familiarize yourself with the various features, like the input box for prompts, output area for responses, and any additional tools available.

Step 7: Write a prompt

Now that you understand the interface, type a prompt into the input box. A prompt is a question or statement that guides Chat-GPT to generate a relevant response. For example, you could ask, "What are some benefits of using solar energy?"

Step 8: Submit the prompt

Press enter or click on the "Submit" button to send your prompt to Chat-GPT. In a few moments, Chat-GPT will generate a response based on your prompt.

Step 9: Review the response

Read the response provided by Chat-GPT. If you are satisfied with the answer, you can save or copy it for future use. If you need more information or a different response, you can try rephrasing your prompt or asking a follow-up question.

STEP 10: Continue using Chat-GPT

You can continue to use Chat-GPT to generate ideas, answer questions, or create content. Just remember to follow any guidelines and limitations set by OpenAI to ensure a positive experience.

As you become more familiar with Chat-GPT, you can explore additional features and tools that may be available with your subscription plan.

PART II

BUILD YOUR EMPIRE

7

GENERATE PROFITABLE IDEAS

Discover how AI can help you brainstorm and refine your business ideas

The ecommerce industry has been good to me—giving me unparalleled freedom and reliable income ever since I graduated college over two decades ago and started my first online store. I am proud to say that I have never had a boss in all these years, nor have I ever seen such an amazing opportunity for other dreamers like me to achieve even greater success—and with far less effort, time, and resources—than I do right now with the latest developments in AI. I am convinced that anyone can start and grow a profitable, future-ready business or brand with Chat-GPT right now, and this book is both a testament to my conviction and a rallying cry to anyone who has ever dreamed of a life outside the rat race.

Starting a business can be both exciting and overwhelming, and it's the latter which has served as a barrier for many people through the years. One of the most challenging is essentially the first piece of the puzzle: the process of generating innovative and viable ideas. But now, CHAT-GPT, an advanced AI language model, offers entrepreneurs a powerful resource for brainstorming and refining business ideas. Let's explore the specific ways in which CHAT-GPT can help you generate ideas to kickstart your entrepreneurial journey.

Identifying Business Niches

A profitable niche is the foundation of a successful business. By providing CHAT-GPT with information about your interests, skills, and industry trends, the AI can generate a list of potential niches for you to consider. This can save you countless hours of research and help you uncover unique opportunities that may have otherwise gone unnoticed.

Assessing Market Demand

Once you've identified a niche, it's crucial to ensure there's sufficient market demand to support your business idea. CHAT-GPT can help you analyze customer reviews, social media conversations, and other data sources to gauge demand for specific products or services. This valuable insight can help you refine your idea and identify potential gaps in the market.

Generating Unique Selling Propositions (USPs)

A strong USP can set your business apart from the competition. CHAT-GPT can help you brainstorm USPs for your product or service by suggesting differentiators, such as innovative features, superior quality, or excellent customer service. The AI can also provide examples of successful USPs from similar businesses, giving you inspiration for your own unique value proposition.

Creating a Brand Identity

A memorable brand identity is essential for connecting with your target audience and building customer loyalty. CHAT-GPT can help you develop your brand's mission, vision, and core values, as well as suggest a creative name, tagline, and logo design. By guiding you through the process of creating a cohesive brand identity, CHAT-GPT ensures that your business stands out in the marketplace.

Developing Business Models

Selecting the right business model is crucial for long-term success. CHAT-GPT can help you explore various business models, such as subscription-based services, e-commerce, or franchising, and analyze their pros and cons. By providing you with a comprehensive understanding of each model, the AI can help you choose the one that best aligns with your business goals and resources.

Identifying Potential Partners and Suppliers

Building a strong network of partners and suppliers is essential for any business. CHAT-GPT can help you identify potential collaborators by suggesting companies, professionals, or suppliers who share your vision and can contribute to your business growth. The AI can even help you draft a pitch to reach out to these potential partners, increasing your chances of forging successful relationships.

Creating a Preliminary Business Plan

A well-crafted business plan is essential for securing funding and guiding your business strategy. CHAT-GPT can help you draft the key sections of your business plan, such as the executive summary, market analysis, and marketing strategy, by providing templates or guiding you through the process. This will ensure you have a solid foundation upon which to build your business.

Supercharged CHAT-GPT Prompts for Idea Generation

To get the best results when using CHAT-GPT for the idea generation process, use the following prompts as a guide:

1. **Identifying Business Niches** | Prompt: "I am interested in technology, sustainability, and e-commerce. Can you suggest potential business niches that combine these interests and have a growing market demand?"
2. **Assessing Market Demand** | Prompt: "I'm considering starting a business that sells eco-friendly, reusable food storage containers. Can you analyze market demand and trends for this product based on customer reviews and social media conversations?"
3. **Generating Unique Selling Propositions (USPs)** | Prompt: "I'm planning to launch an online fitness coaching service. Can you suggest potential USPs that would differentiate my business from other fitness coaching services in the market?"
4. **Creating a Brand Identity** | Prompt: "I want to create a brand identity for my sustainable clothing line, focused on ethically sourced materials and eco-friendly production. Can you help me develop the brand's mission, vision, core values, name, and tagline?"
5. **Developing Business Models** | Prompt: "I'm considering starting a meal delivery service that offers healthy and affordable meals. Can you provide an analysis of various business models, such as subscription-based, pay-per-meal, or a hybrid model, with their pros and cons?"
6. **Identifying Potential Partners and Suppliers** | Prompt: "I'm looking for potential partners and suppliers for my eco-friendly beauty products business. Can you suggest companies or professionals with expertise in sustainable packaging, natural ingredients sourcing, and ethical manufacturing?"

7. **Creating a Preliminary Business Plan** | Prompt: "I want to create a preliminary business plan for my mobile app development company, focusing on custom solutions for small businesses. Can you guide me through the process of writing key sections like the executive summary, market analysis, and marketing strategy?"

Remember to be as specific as possible when crafting your prompts, as this will help CHAT-GPT generate more accurate and relevant results for your business idea generation needs.

8

INDUSTRY IDEAS FOR CHAT-GPT

Artificial intelligence (AI) can be customized and adapted to fit the unique requirements of various types of businesses. By utilizing AI's capabilities, businesses can streamline processes, enhance customer experiences, and optimize operations. Here are examples of how AI can assist different types of businesses:

Retail and Ecommerce:

- Personalized product recommendations
- AI-powered chatbots for customer service
- Inventory and supply chain management
- Visual search and virtual fitting rooms

Healthcare:

- Medical image analysis and diagnostics
- Drug discovery and development
- Virtual health assistants and telemedicine

- Predictive analytics for patient care

Finance:

- Fraud detection and prevention
- Credit risk assessment and underwriting
- Robo-advisors for investment management
- Algorithmic trading and financial forecasting

Manufacturing:

- Predictive maintenance for equipment
- Quality control and defect detection
- Optimized production planning and scheduling
- Collaborative robots (cobots) for automation

Agriculture:

- Precision farming through data analysis
- Crop health monitoring using drones
- Automated irrigation and fertilization systems
- Yield prediction and pest control

Transportation and Logistics:

- Route optimization for delivery vehicles
- Autonomous vehicles and drones
- Real-time traffic analysis and prediction
- Demand forecasting and fleet management

Marketing and Advertising:

- AI-driven content generation and curation
- Programmatic advertising and media buying
- Customer segmentation and targeting
- Sentiment analysis and social listening

Human Resources:

- Resume screening and candidate matching
- AI-powered interview scheduling
- Employee engagement and sentiment analysis
- Predictive analytics for talent retention

Almost any type of business can benefit from AI in different ways based on its specific needs and industry trends. By giving Chat-GPT a prompt that lists a specific industry, it will generate similar ideas for just about anything.

For example, try a prompt like: *"Give examples of different ways in which AI can assist the ____ industry."*

Chat-GPT can likewise provide practical, general, and even inspirational advice about starting a business outside the context of AI. For example, when I asked Chat-GPT to tell me its top ten business tips for entrepreneurs, the article it generated serves as an excellent foundation from which to venture deeper into the specifics of starting a business with the help of AI:

Business Tips for Entrepreneurs

Starting a business is an exciting journey filled with both opportunities and challenges. As an entrepreneur, knowing how to navigate these obstacles is crucial for success. Here are ten business tips to help you maximize your potential and achieve your entrepreneurial goals.

1. **Have a Clear Vision:** Before starting a business, it is essential to have a clear vision of what you want to achieve. Define your mission, set achievable goals, and create a roadmap to guide your journey. This clarity will help you stay focused and make informed decisions as your business evolves.
2. **Conduct Thorough Market Research:** Understand your target market and industry by conducting thorough research. Identify potential customers, analyze competitors, and examine market trends. This information will help you create a product or service that meets the needs of your audience and stands out in the market.
3. **Develop a Robust Business Plan:** A well-crafted business plan is the foundation of your entrepreneurial journey. It should include an executive summary, market analysis, company description, organization and management structure, marketing and sales strategies, financial projections, and funding requirements. Regularly revisit and update your plan as your business grows.
4. **Build a Strong Network:** Connect with other entrepreneurs, industry experts, and potential partners to expand your network. Attend networking events, join online forums, and engage with others on social media platforms. Building relationships can lead to valuable insights, new opportunities, and increased visibility for your business.
5. **Focus on Customer Satisfaction:** Your customers are the lifeblood of your business. Prioritize their needs and provide exceptional customer service. Seek feedback, listen to their concerns, and continuously improve your product or service based on their input. Satisfied customers are more likely to become repeat clients and recommend your business to others.

6. **Manage Your Finances Wisely:** Effective financial management is essential for business success. Track your expenses, monitor cash flow, and create realistic budgets. Make informed financial decisions and seek professional advice when needed. Proper financial management will help you avoid potential pitfalls and ensure long-term stability.
7. **Embrace Digital Marketing:** In today's digital age, having a strong online presence is vital. Develop a professional website, engage with your audience on social media, and invest in search engine optimization (SEO). Utilize email marketing and content creation to drive traffic to your site and convert visitors into customers.
8. **Be Adaptable:** The business world is constantly changing, and adaptability is key to success. Stay informed about industry trends, be open to new ideas, and be prepared to pivot your strategy when necessary. Embracing change can help you stay ahead of the competition and capitalize on new opportunities.
9. **Outsource Strategically:** As an entrepreneur, it's impossible to be an expert in every aspect of your business. Outsource tasks that are outside your core competencies, such as accounting or graphic design. This will free up your time to focus on your strengths and strategic decision-making.
10. **Persevere and Learn from Failure:** Starting a business is not a linear path, and setbacks are inevitable. Embrace failure as an opportunity to learn and grow. Maintain a positive mindset, stay persistent, and keep refining your strategies until you achieve the success you desire.

By implementing these business tips, you'll be well on your way to entrepreneurial success, which can be amplified exponentially with the help of Chat-GPT and other AI tools that are available today. Stay focused on your vision, embrace challenges, and continuously adapt and refine your strategies. With hard work, perseverance, and a little bit of luck, you can turn your dream business into a reality.

9

BUSINESS PLANNING FOR STARTUPS

Utilize AI to craft a comprehensive and effective business plan for your startup

A well-crafted business plan is crucial for the success of any startup. It serves as a roadmap that outlines your objectives, strategies, and financial projections, guiding your business growth and helping secure funding. Chat-GPT, an advanced AI language model, can be a valuable resource in the business planning process, offering specific guidance and insights. Let's explore the ways in which Chat-GPT can assist you in creating a robust business plan, along with example prompts for each aspect.

Executive Summary

The executive summary provides an overview of your business plan and should capture the essence of your startup.

Example Prompt: "I'm starting a sustainable fashion brand that focuses on ethically sourced materials and eco-friendly production.

Can you help me draft a compelling executive summary for my business plan?"

Company Description

This section describes your company's mission, vision, and values, as well as the problem it solves or the need it addresses.

Example Prompt: "My company will provide an online platform for local artisans to sell their handcrafted products. Can you help me write a company description that highlights our mission, vision, and the value we bring to the market?"

Market Analysis

A thorough market analysis demonstrates your understanding of the industry, competition, and target audience.

Example Prompt: "I plan to launch a vegan meal delivery service in San Francisco. Can you help me conduct a market analysis, focusing on the target audience, market size, trends, and key competitors?"

Marketing and Sales Strategies

Outlining your marketing and sales strategies will showcase how you plan to attract and retain customers.

Example Prompt: "My business will offer online fitness coaching through a mobile app. Can you help me develop marketing and sales strategies to reach and engage my target audience, including advertising channels, pricing, and promotional tactics?"

Organizational Structure

Describing your organizational structure and management team highlights the expertise and experience behind your startup.

Example Prompt: "My startup is developing an AI-powered customer support platform. Can you help me outline our organizational structure, including key roles and responsibilities, and describe the expertise of our management team?"

Product Line or Service Description

This section provides detailed information about your products or services and their unique selling points.

Example Prompt: "I'm starting a business that sells 3D-printed, customizable home décor items. Can you help me describe our product line, emphasizing the customization options, quality, and sustainability aspects?"

Financial Projections

Financial projections help investors understand the potential profitability and growth of your business.

Example Prompt: "I plan to open a boutique bakery specializing in artisanal pastries and bread. Can you help me create a three-year financial projection, including revenue, expenses, and breakeven analysis?"

Creating a comprehensive and effective business plan is an essential step in the entrepreneurial journey. Chat-GPT offers invaluable support in crafting each section of your plan, providing insights and guidance that can bolster your business's potential for success.

10

CONTENT CREATION FOR STARTUPS

Leverage AI to create compelling content and drive your business growth

Content creation and copywriting are essential for promoting your startup and attracting potential customers. As a new business owner, you may find it challenging to produce high-quality content consistently. Chat-GPT, an AI-powered language model, can be an invaluable resource in generating engaging content across various platforms. Let's explore specific ways to use Chat-GPT for content creation and copywriting and provide example prompts to help you get started.

Website Content and Product Descriptions

Your website is the digital storefront for your business, and compelling content is crucial for capturing your audience's attention. Chat-GPT can help you create engaging website content, including product descriptions that showcase the unique value of your offerings.

Example Prompt: "I am launching an online store selling eco-friendly activewear. Can you help me write a captivating homepage content and a product description for a sustainable yoga mat?"

Blog Posts and Articles

Blogging is an excellent way to establish your brand's authority, engage with your audience, and improve your search engine rankings. Chat-GPT can help you generate relevant blog post ideas or even write entire articles on your chosen topics.

Example Prompt: "I run a nutrition-focused startup. Can you help me write a blog post about the benefits of incorporating plant-based protein sources into our diet?"

Social Media Content

Creating engaging social media content can be time-consuming. Chat-GPT can help you craft compelling copy for various platforms like Facebook, Instagram, Twitter, and LinkedIn, tailored to your brand's voice and target audience.

Example Prompt: "My startup offers personalized meal plans. Can you help me write a series of Instagram captions to promote our meal plans and encourage sign-ups?"

Email Campaigns and Newsletters

Email marketing is an effective way to reach your audience, promote your products, and nurture customer relationships. Chat-GPT can help you write captivating subject lines, email copy, and newsletter content that encourages opens, clicks, and conversions.

Example Prompt: "I am launching a Kickstarter campaign for my innovative smartwatch startup. Can you help me craft a persuasive email to announce the launch and encourage backers to support our project?"

Press Releases and Media Pitches

Attracting media coverage can boost your startup's visibility and credibility. Chat-GPT can help you write professional press releases and persuasive media pitches that effectively communicate your news and pique the interest of journalists and influencers.

Example Prompt: "My tech startup has developed a revolutionary wearable fitness tracker. Can you help me write a press release announcing the product launch and highlighting its unique features?"

Ad Copy and Promotional Materials

Crafting effective ad copy and promotional materials is essential for driving interest and conversions. Chat-GPT can help you generate persuasive copy for digital and print advertisements, brochures, flyers, and other promotional materials.

Example Prompt: "My startup offers a cloud-based project management tool. Can you help me write compelling ad copy for a Google Ads campaign targeting small business owners?"

Video Scripts and Podcast Outlines

Engaging multimedia content, such as videos and podcasts, can help you reach a wider audience and enhance your brand's presence. Chat-GPT can assist you in creating captivating video scripts and podcast outlines that resonate with your target market.

Example Prompt: "I'm starting a podcast for my travel planning startup. Can you help me outline a podcast episode about the top 10 hidden gems to explore in Europe?"

Chat-GPT is a powerful tool that can help you streamline your content creation and copywriting efforts when starting a business. By utilizing AI, you can generate high-quality, engaging content that captures your audience's attention and drives business growth. With the help of Chat-GPT, you can overcome the challenges of content creation and focus on other crucial aspects of your startup journey.

More AI Tools for Startups

AI tools like the ones listed below have revolutionized marketing, design, content creation, and a variety of other tasks for business. These tools and others like them reduce the time and effort required to create high-quality content. As AI technology continues to evolve, we can expect to see even more innovative and useful tools emerge in the near future.

1. Copy.ai: AI tool for generating copy that increases conversions. With its advanced AI algorithms, it can write sales pages, blog posts, ads, and more in minutes. https://www.copy.ai/
2. CopyMonkey: AI tool for creating Amazon listings in seconds. It offers a suite of tools to automate your Amazon business, from keyword research to copywriting. http://copymonkey.ai/
3. Ocoya: AI tool for creating and scheduling social media content. It helps you plan and publish your social media posts effortlessly, so you can focus on what really matters: engaging with your audience. https://www.ocoya.com/
4. Inkforall: AI tool for content generation, optimization, and performance. It offers a suite of tools that help you write high-quality content faster, optimize your content for search engines and measure its performance so that you can make data-driven decisions. https://inkforall.com/
5. Thundercontent: AI tool for generating content. Using its advanced language functionality and machine learning algorithms, it provides a platform for

generating high-quality content, from blog posts to articles, in minutes. https://thundercontent.com/

By now you're probably beginning to see the vast potential of AI to create content on an epic scale, elevate your brand's presence, connect with your target audience, and set your business on a path to success.

11

SOCIAL MEDIA MANAGEMENT FOR STARTUPS

Leverage AI to create a consistent and engaging social media presence for your new business

Effective social media management is crucial for startups looking to increase brand awareness, engage with their target audience, and drive sales. Chat-GPT, an AI-powered language model, can be an invaluable resource for managing your social media presence and generating compelling content tailored to your brand's voice and audience. Let's explore specific ways to use Chat-GPT for social media management when starting a business, along with example prompts to help you get started.

Identifying the Right Social Media Platforms

Choosing the right social media platforms is essential for reaching your target audience. Chat-GPT can help you evaluate the pros and cons of each platform based on your business type and target audience.

Example Prompt: "I am starting an online store for handmade candles. Can you help me decide which social media platforms would be most effective for reaching my target audience and promoting my products?"

Creating a Social Media Strategy

A well-crafted social media strategy is key to achieving your marketing goals. Chat-GPT can help you develop a comprehensive strategy that outlines your objectives, content mix, and posting frequency.

Example Prompt: "I am starting a tech consulting firm targeting small businesses. Can you help me create a social media strategy that focuses on showcasing our expertise and driving leads?"

Crafting Engaging Social Media Content

Consistently creating captivating social media content can be time-consuming. Chat-GPT can help you generate engaging copy for various platforms, tailored to your brand voice and target audience.

Example Prompt: "My startup offers subscription-based meal kits. Can you help me write a series of tweets that promote our meal kits and highlight the benefits of using our service?"

Curating Relevant Content to Share

Sharing third-party content relevant to your industry can keep your audience engaged and position your brand as a thought leader. Chat-GPT can help you identify interesting articles, videos, and other content to share on your social media channels.

Example Prompt: "My startup focuses on sustainable fashion. Can you help me find five articles about the latest eco-friendly fashion trends and advancements to share on our Facebook page?"

Generating Social Media Post Ideas

Coming up with fresh social media post ideas can be challenging. Chat-GPT can help you brainstorm creative and diverse post ideas that resonate with your target audience and align with your brand values.

Example Prompt: "My startup offers online coding courses for kids. Can you help me generate ten Instagram post ideas that showcase the benefits of learning to code at a young age?"

Responding to Comments and Messages

Engaging with your audience through comments and direct messages is vital for building relationships and fostering customer loyalty. Chat-GPT can help you draft thoughtful responses to various inquiries, comments, and feedback.

Example Prompt: "A customer on our Facebook page asked about the ingredients in our skincare products. Can you help me write a response that addresses their concerns and highlights our commitment to using natural ingredients?"

Analyzing Social Media Performance

Monitoring and analyzing your social media performance is crucial for identifying what works and what needs improvement. Chat-GPT can help you review your social media metrics and provide insights and recommendations for optimizing your strategy.

Example Prompt: "I've been running a LinkedIn campaign for my B2B software startup for the past month. Can you help me analyze the performance of our campaign and suggest areas for improvement?"

Embracing the power of Chat-GPT for social media management can help you create a consistent, engaging, and effective social media presence for your startup. By leveraging AI-generated content and insights, you can save time and resources while connecting with your target audience and creating a thriving business.

12

MARKET RESEARCH

Harness the power of AI to identify opportunities, assess competition, and understand your target audience

Market research is a crucial step in starting a business, as it helps entrepreneurs identify opportunities, assess competition, and understand their target audience. With Chat-GPT, a powerful AI language model, businesses can gain valuable insights and streamline their market research efforts. Let's explore specific ways to use Chat-GPT for market research when starting a business, along with example prompts to help you get started.

Identifying Market Opportunities

Chat-GPT can help you identify market opportunities by analyzing trends, customer needs, and potential gaps in the market. By utilizing AI-generated insights, you can make data-driven decisions when defining your product or service offering.

Example Prompt: "I'm starting a vegan snack subscription box service. Can you help me identify market opportunities and customer needs in the vegan snack industry?"

Assessing Competition

Understanding your competition is essential for identifying your unique selling proposition (USP) and positioning your business effectively. Chat-GPT can help you analyze your competitors and their strategies to provide insights for refining your own approach.

Example Prompt: "My online clothing store will specialize in sustainable and ethically-produced apparel. Can you help me analyze my top three competitors and suggest ways we can differentiate ourselves and improve our strategy?"

Understanding Your Target Audience

Knowing your target audience's preferences, pain points, and needs is vital for creating a successful product or service. Chat-GPT can help you develop detailed buyer personas and identify the most effective ways to reach and engage with your target audience.

Example Prompt: "My business will offer a meal planning and grocery delivery service for busy professionals. Can you help me create a detailed buyer persona and suggest ways to reach and engage with this target audience?"

Evaluating Market Size and Growth Potential

Estimating the size and growth potential of your target market can help you understand the potential demand for your product or service. Chat-GPT can assist you in evaluating market size, growth rates, and trends to inform your business decisions.

Example Prompt: "I'm planning to open a coworking space in Austin, Texas. Can you help me estimate the market size and growth potential for coworking spaces in that area?"

Analyzing Industry Trends

Staying informed about industry trends is crucial for positioning your business for success. Chat-GPT can help you identify and analyze key trends in your industry, providing insights into potential opportunities and challenges.

Example Prompt: "My business will develop and sell eco-friendly beauty products. Can you help me identify and analyze the key trends in the sustainable beauty industry?"

Pricing Strategy Development

Developing an effective pricing strategy is essential for ensuring profitability and competitiveness. Chat-GPT can help you analyze pricing strategies of competitors, market conditions, and customer expectations to develop a pricing model that works for your business.

Example Prompt: "I'm starting a subscription-based online fitness platform. Can you help me analyze the pricing strategies of similar platforms and suggest a pricing model that balances competitiveness and profitability?"

Conducting a SWOT Analysis

A SWOT analysis (Strengths, Weaknesses, Opportunities, Threats) can provide valuable insights into your business's internal and external factors. Chat-GPT can help you conduct a comprehensive SWOT analysis to identify areas for improvement and capitalize on opportunities.

Example Prompt: "I'm opening a boutique hotel in a popular tourist destination. Can you help me conduct a SWOT analysis for my business to identify our strengths, weaknesses, opportunities, and threats?"

By leveraging Chat-GPT for market research when starting a

business, entrepreneurs can gain valuable insights, identify opportunities, and make informed decisions.

13

PRODUCT DEVELOPMENT

Harness the power of AI to develop lucrative products

When starting a business, developing a product that meets market needs and stands out from the competition is vital. Chat-GPT, an advanced AI language model, can help entrepreneurs generate ideas, refine product designs, and optimize their offerings. Let's explore specific ways to use Chat-GPT for product development when starting a business, along with example prompts to help you get started.

Idea Generation

Chat-GPT can help you brainstorm innovative product ideas by providing suggestions based on your industry, target audience, and unique selling points.

Example Prompt: "I'm starting a business in the sustainable fashion industry. Can you help me generate unique and innovative product ideas that cater to eco-conscious consumers?"

Identifying Key Features

Determining the right features for your product can make all the difference in its success. Chat-GPT can assist you in identifying the most important and desired features for your target audience.

Example Prompt: "I'm developing a smart home security system. Can you help me identify the most important features and functionalities that customers are looking for in a home security system?"

Refining Product Design

A well-designed product can improve user experience and increase customer satisfaction. Chat-GPT can provide design suggestions and improvements based on industry best practices and user preferences.

Example Prompt: "I'm designing a fitness tracker for seniors. Can you help me refine the design to make it more user-friendly and appealing to the senior demographic?"

Evaluating and Improving Product Concepts

Critiquing and refining your product concept is essential to ensure it meets market needs and expectations. Chat-GPT can help you evaluate your product concept and provide recommendations for improvements.

Example Prompt: "My product concept is a reusable water bottle with a built-in water purification system. Can you help me evaluate the concept and suggest potential improvements?"

Identifying Potential Challenges and Risks

Being aware of potential challenges and risks associated with your product can help you mitigate them and develop a more robust offering. Chat-GPT can help you identify these challenges and suggest strategies to overcome them.

Example Prompt: "I'm planning to launch a line of organic baby food products. Can you help me identify potential challenges and risks in this market and suggest strategies to mitigate them?"

Conducting a Competitive Analysis

Understanding your competition is crucial for developing a product that stands out. Chat-GPT can help you analyze competing products and identify areas where you can differentiate and excel.

Example Prompt: "My business will offer an online language learning platform. Can you help me analyze the top three competitors in the market and suggest ways we can differentiate our platform and provide a unique value proposition?"

Assessing Market Viability

Determining the market viability of your product is key to ensuring its success. Chat-GPT can help you assess the market demand, potential competition, and pricing strategies for your product.

Example Prompt: "I want to start a business selling vegan leather handbags. Can you help me assess the market viability of this product, including demand, competition, and pricing considerations?"

Chat-GPT can be a valuable resource for entrepreneurs looking to develop innovative products and stay ahead of the competition. By utilizing AI-generated insights and suggestions, you can optimize your product development process, making data-driven decisions, and ensuring the success of your new business. Embrace the potential of AI and give your business a competitive edge with the help of Chat-GPT.

14

CUSTOMER SUPPORT

No matter what type of business you start, one of the best ways to gain repeat customers and word-of-mouth advertising—which, in turn, leads to long-term growth and profitability—is by offering exceptional customer support. There are a number of excellent customer support systems on the market, many of which already integrate with AI in some form. Here is a list of the most popular, with Pros and Cons for each:

Zendesk

Pros:

- Intuitive and user-friendly interface
- Multichannel support (email, live chat, phone, social media)
- Scalable plans for businesses of all sizes
- Strong integration capabilities with popular ecommerce platforms and CRMs

Cons:

- Can be expensive for smaller businesses
- Customization options may require technical expertise

Freshdesk

Pros:

- Affordable pricing with a free plan for small businesses
- Built-in live chat and phone support
- Easy to set up and use
- Integrates with popular ecommerce platforms and CRMs

Cons:

- Limited reporting capabilities on lower-tier plans
- May require add-ons for advanced features

Help Scout

Pros:

- Simple, email-based ticketing system
- Collaborative features for team communication
- Integrates with popular ecommerce platforms and other tools
- Customizable with a range of add-ons and integrations

Cons:

- No built-in live chat or phone support (requires integration)
- Pricing may be expensive for smaller businesses

Intercom

Pros:

- Unified customer support and communication platform
- Powerful live chat and chatbot capabilities
- Customizable messaging and automation features
- Supports integrations with popular ecommerce platforms

Cons:

- Can be expensive, particularly for larger teams
- Steeper learning curve compared to other options

LiveAgent

Pros:

- Affordable pricing with a wide range of features
- Native live chat, phone support, and video chat capabilities
- Extensive reporting and analytics
- Integrates with popular ecommerce platforms and CRMs

Cons:

- User interface may feel dated compared to competitors
- Customization options may be limited

When selecting the best customer support software for your online store, consider factors such as pricing, features, ease of use, integration capabilities, and scalability to ensure that the solution meets your business's unique needs.

Personally, I recommend Zendesk because of its ease-of-use and exceptional features, and I have used it for many years for my online stores. Nowadays, AI is integrated into Zendesk through various features and tools that help enhance customer support experiences, streamline workflows, and improve overall efficiency.

Some of the AI-powered features in Zendesk include:

1. **Answer Bot:** The Answer Bot is an AI-driven chatbot designed to provide automated responses to customer inquiries by using machine learning algorithms to search through your knowledge base articles. This helps resolve customer queries faster, reduces support team workload, and increases customer satisfaction.
2. **Automatic Ticket Routing:** Zendesk uses AI to analyze incoming support requests and automatically route them to the appropriate support team or agent based on factors such as skills, availability, and workload. This helps ensure that tickets are handled by the most suitable agent and reduces response times.
3. **Content Cues:** Content Cues is an AI-powered feature that analyzes customer support interactions to identify patterns and trends in customer inquiries. It provides insights and suggestions for improving your knowledge base content, making it more relevant and helpful to customers.
4. **Machine Learning-Based Analytics:** Zendesk's analytics tools use machine learning algorithms to analyze support data and identify trends, patterns, and areas for improvement. This helps businesses make data-driven decisions to enhance their customer support operations.
5. **Natural Language Processing (NLP):** Zendesk's AI capabilities also include NLP, which helps understand and process human language in customer support requests. This enables better ticket categorization,

sentiment analysis, and more effective automation of support processes.

These AI-powered features and tools in Zendesk allow businesses to deliver more efficient and personalized customer support, leading to increased customer satisfaction and loyalty. Plus, you can be assured that Zendesk and other support software will continue to find news ways of employing the power of AI into its functionality.

15

BRANDING FOR STARTUPS

Leverage AI to build a strong personal brand and collaborate with influencers for maximum impact

In today's digital age, personal branding and influencer marketing play a significant role in the success of a new business. Chat-GPT, an advanced AI language model, can help entrepreneurs build a strong personal brand and collaborate with influencers effectively. Let's explore specific ways to use Chat-GPT for personal branding and influencer marketing when starting a business, along with example prompts to help you get started.

Defining Your Personal Brand

Your personal brand is a reflection of your values, expertise, and unique selling points. Chat-GPT can help you articulate your personal brand effectively, ensuring that it resonates with your target audience.

Example Prompt: "I'm a sustainability expert starting an eco-friendly consulting business. Can you help me define my personal brand to appeal to environmentally conscious clients and organizations?"

Crafting a Compelling Brand Story

A strong brand story can help you connect with your audience on a deeper level. Chat-GPT can assist you in developing a compelling narrative that highlights your journey, expertise, and values.

Example Prompt: "I'm a fitness coach specializing in helping busy professionals stay fit. Can you help me craft a compelling brand story that showcases my passion for fitness and my ability to help clients achieve their goals?"

Creating Engaging Content

High-quality, engaging content is crucial for building your personal brand and attracting an audience. Chat-GPT can generate content ideas and even draft blog posts, social media updates, or video scripts that align with your brand.

Example Prompt: "I'm a financial advisor for young professionals, and I need content ideas for my blog. Can you suggest engaging topics that will resonate with my target audience and showcase my expertise?"

Identifying Influencers for Collaboration

Collaborating with influencers can amplify your reach and lend credibility to your personal brand. Chat-GPT can help you identify potential influencers within your niche who align with your brand values and goals.

Example Prompt: "My business focuses on cruelty-free makeup products. Can you help me identify influencers in the beauty industry who share similar values and might be interested in collaborating with my brand?"

Crafting Influencer Outreach Messages

Reaching out to influencers requires a personalized and compelling message. Chat-GPT can help you draft tailored outreach messages that showcase your brand and highlight potential collaboration opportunities.

Example Prompt: "I'm launching a plant-based meal delivery service and would like to collaborate with a health and wellness influencer. Can you help me draft an outreach message that highlights the benefits of partnering with my brand?"

Developing Influencer Marketing Campaign Ideas

Effective influencer marketing campaigns can drive brand awareness and conversions. Chat-GPT can help you generate creative campaign ideas that leverage the unique strengths of your chosen influencers.

Example Prompt: "I've partnered with a travel influencer for my eco-friendly travel gear brand. Can you help me come up with a creative influencer marketing campaign that promotes our products and resonates with their audience?"

Evaluating Influencer Marketing Results

Measuring the success of your influencer marketing efforts is crucial for optimizing future campaigns. Chat-GPT can help you analyze campaign results and suggest improvements for future collaborations.

Example Prompt: "I recently ran an influencer marketing campaign for my online course platform. Can you help me evaluate the results and suggest ways to improve our influencer marketing efforts moving forward?"

By harnessing the power of Chat-GPT for personal branding and influencer marketing, entrepreneurs can build a strong presence, engage their target audience, and drive business growth. The future of influencers may very well belong to those who embrace the potential of AI.

16

CREATE A MOBILE APP

Mobile apps have become an essential part of our digital lives. From managing daily tasks to staying connected with friends, apps make our lives easier and more convenient. Creating a mobile app can be a profitable venture whether it serves as an extension or utility for a new startup or existing business, or as a stand-alone mobile app that generates income on its own.

If you're an entrepreneur or developer looking to create a mobile app, Chat-GPT can be a valuable resource. Let's explore how you can leverage Chat-GPT to assist in various stages of mobile app development.

Idea Generation and Validation

Before diving into app development, it's important to have a clear idea of what your app will do and how it will benefit users. Chat-GPT can help you brainstorm app ideas and validate their potential. You can ask Chat-GPT questions like, "What are some unique app ideas for improving productivity?" or "How can I differentiate my app idea from similar apps in the market?"

Planning and Design

Once you have a solid app idea, it's time to plan and design the app. Chat-GPT can help you create user stories, define app features, and outline the overall user experience. You can ask Chat-GPT questions like, "What are the key features of a fitness tracking app?" or "How can I create a user-friendly interface for my budgeting app?"

Wireframing and Prototyping

Before coding the app, it's essential to create wireframes and prototypes to visualize the app's layout and functionality. Chat-GPT can provide guidance and suggestions on creating wireframes, user flows, and interactive prototypes. You can ask questions like, "What are the best practices for wireframing a mobile app?" or "How can I create an interactive prototype for my travel planning app?"

Coding Assistance

While Chat-GPT may not write entire chunks of code for your app, it can still assist you in solving specific coding problems or providing guidance on best practices. You can ask Chat-GPT questions like, "What's the best way to implement push notifications in a React Native app?" or "How can I optimize the performance of my mobile app?"

Testing and Debugging

Testing and debugging are essential steps in app development to ensure a smooth user experience. Chat-GPT can help you identify common issues, suggest testing methodologies, and provide solutions for specific problems. You can ask questions like, "What are some common bugs in mobile apps and how can I fix them?" or "What are the best practices for testing mobile app performance?"

Marketing and Promotion

Once your app is developed, it's time to promote it and attract users. Chat-GPT can help you develop marketing strategies, craft compelling app store descriptions, and create engaging promotional content. You can ask questions like, "What are the most effective ways to market a mobile app?" or "How can I write an engaging app store description for my productivity app?"

User Feedback and Iteration

To ensure your app's success, it's essential to gather user feedback and make improvements based on their suggestions. Chat-GPT can help you analyze user feedback, prioritize improvements, and plan updates. You can ask questions like, "How can I effectively gather user feedback for my app?" or "What features should I prioritize in my app's next update?"

Enlisting the “expertise” of Chat-GPT can significantly enhance the app development process. From generating app ideas to marketing and promotion, Chat-GPT can provide valuable insights and assistance every step of the way. By harnessing its capabilities, you can create a successful mobile app that meets the needs of your target audience and stands out in the competitive app market.

17

AI POWER TOOLS

The power of artificial intelligence, particularly Chat-GPT, is transforming the way we interact with technology and access information. By leveraging Chat-GPT, entrepreneurs and developers can create innovative online tools that deliver immense value to users. In this chapter, we'll discuss the process of creating an AI-based Chat-GPT-powered online tool and present the top five ideas for such tools.

1. **Defining the Purpose and Target Audience**

Before developing a Chat-GPT-powered online tool, it's crucial to identify the purpose it will serve and the target audience. Consider the problems your tool will solve, the value it provides, and the demographics it will cater to.

1. **Creating the User Interface (UI)**

Design a user-friendly interface that is easy to navigate and visually appealing. The UI should facilitate seamless interaction between users and the Chat-GPT-powered tool, ensuring a satisfying user experience.

1. **Integrating Chat-GPT into Your Tool**

Connect your tool with the Chat-GPT API, allowing it to access the AI's capabilities. Ensure that the API calls are optimized to minimize latency and provide prompt responses to user queries.

1. **Testing and Iterating**

Test your Chat-GPT-powered tool rigorously to identify any bugs or areas that need improvement. Gather feedback from users and iterate on your tool to enhance its performance and usability.

Top 5 Chat-GPT-Powered Online Tool Ideas

1. **Content Generation Tool**

Create a tool that generates high-quality content for blog posts, social media updates, and marketing materials. Users can provide keywords, topics, or prompts, and the tool will generate relevant and engaging content using Chat-GPT's natural language processing capabilities.

1. **SEO Optimization Assistant**

Develop an online tool that helps users optimize their website content for search engines. The tool can analyze users' content, suggest relevant keywords, and provide tips on improving readability and user engagement.

1. **Language Learning Companion**

Create an interactive language learning tool that allows users to practice their conversational skills with a Chat-GPT-powered AI tutor. The tool can provide real-time feedback on grammar, pronunciation, and vocabulary, and engage users in immersive, context-driven conversations.

1. **Coding Problem Solver**

Develop a Chat-GPT-powered tool that assists users in solving coding problems and understanding programming concepts. Users can input their questions or issues, and the tool can provide guidance, solutions, or suggestions on best practices.

1. **Personalized Career Advisor**

Create an online tool that offers personalized career advice based on users' skills, interests, and goals. The Chat-GPT-powered tool can recommend suitable career paths, provide insights on industry trends, and suggest resources for skill development and networking.

My Own Real-World Example

For an easy platform to build your first Chat-GPT online tool, I recommend creating a simple Wordpress website and using a premade Chat-GPT plugin or asking Chat-GPT to create one for you. I explore this more in the Blogging section of this book, but to get started quickly I recommend the Wordpress plugin called *"AI Engine: ChatGPT Chatbot, Content Generator, GPT 3 & 4, Ultra-Customizable (Pro Version),"* which is what I used to build Versify.AI, an online AI-powered poetry generating website. You can find the plugin by searching in the Plugins section of your Wordpress dashboard.

But first, if you wish to register a .AI domain to use for your Chat-GPT-powered online tool website, my preference is *Name-*

Cheap.com. They also offer very affordable Wordpress hosting and a Wordpress website builder, which makes it even simpler to start creating your AI-powered website quickly and with little tech experience. I used all of these to build Versify.ai.

VERSIFY.AI ~ *Collaborate with AI to create unique poetry based on your poetic preferences, favorite poets, preferred styles, and more!*

Within a few hours of registering my AI domain and buying a $4-per-month hosting package, I was able to create a website that allows users to choose which poem form they wish to create (such as sonnet, ode, limerick, haiku, or free form) and which poet's style they want to emulate. Then, after the user inputs a subject for their poem and hits the submit button, they are presented with Chat-GPT-generated poem that uses all their preferences. For example, one user asked Versify.ai to create a Haiku poem about robots, and it generated this:

Robots, made of steel,
Silently move through the night,
Their circuits hum soft.

A Haiku poem, in case you don't know, is always structured in three lines: five syllables, seven syllables, and again five syllables. Given I'm not a programmer, I never could have coded a poem generator that had the ability to produce one line of text, let alone a Haiku with its precise structure. But yet, Chat-GPT was able to help me build a tool that it was then able to work through to generate a perfectly-formed Haiku. And, in my experience, I've never heard robots described so beautifully.

If nothing else, this serves as an example of what Chat-GPT truly is at its core: **an advanced language model.**

After being so impressed by Chat-GPT's poetry, I also added a similar function for *Versify.ai* where it generates song lyrics based on the styles of famous musicians, from Bob Dylan to Taylor Swift, and about whatever subject the user specifies. I plan to add additional tools for generating various forms of language in verse form as well.

The next online tools I plan to build using Wordpress include the following:

Paintify.ai — Turn photos, ideas, or sketches into works of art with AI

Typify.ai — The easy AI-powered content writer

Dewy.ai — AI-generated beauty advice based your photo. Free compliments, too!

Spox.ai — Create your own virtual AI-powered spokesperson

Creating a Chat-GPT-powered online tool opens up a world of possibilities for entrepreneurs and developers. By harnessing the power of AI, you can create innovative solutions that cater to the diverse needs of users across various domains. By combining a user-centric design with the cutting-edge capabilities of Chat-GPT, you can develop online tools that have a tangible impact on users' lives and deliver immense value.

18

ONLINE SELLING SUCCESS

When it comes to making money with the help of Chat-GPT, ecommerce is perhaps the fastest and most direct means of generating income. With the wide range of software available on the market now, much of it simplified to the point that little to no coding experience is necessary, starting an online store has never been easier.

One key to ecommerce success is finding a profitable niche, ideally one in which you have experience, even if you merely have an interest in a certain type of product or industry at the moment.

Here some of the most lucrative ecommerce businesses that anyone can start now.

1. **Online Clothing Boutique:** Tap into the ever-growing demand for fashionable clothing and accessories by starting an online boutique. You can specialize in a particular niche, such as sustainable fashion, activewear, or plus-size clothing, to cater to specific market segments.
2. **Customized and Personalized Products:** Offer customized and personalized items such as T-shirts,

mugs, phone cases, or home decor. With print-on-demand services, you can create unique products without investing in inventory upfront.

3. **Health and Wellness Products:** Capitalize on the rising interest in health and wellness by selling supplements, organic products, fitness equipment, or wellness subscriptions.
4. **Eco-friendly and Sustainable Products:** As consumers become more environmentally conscious, there is a growing demand for eco-friendly and sustainable products like reusable bags, bamboo toothbrushes, and solar-powered gadgets.
5. **Subscription Boxes:** Curate and deliver themed subscription boxes to customers on a monthly or quarterly basis. You can target various niches, such as beauty, gourmet food, or hobby supplies.
6. **Online Learning Platforms and Courses:** Leverage your expertise or partner with subject matter experts to create and sell online courses or develop a learning platform on topics like coding, digital marketing, or photography.
7. **Digital Products:** Create and sell digital products like eBooks, templates, stock photos, or software solutions, which require minimal overhead and can be easily distributed online.
8. **Gourmet Food and Specialty Items:** Cater to food enthusiasts by offering gourmet food items, artisanal products, or specialty ingredients that are not readily available in local stores.
9. **Smart Home Gadgets and IoT Devices:** As technology advances, there is a growing demand for smart home gadgets and IoT devices that make daily life more convenient, such as smart speakers, security systems, and energy-efficient lighting.

10. **Dropshipping:** Start a low-investment e-commerce store by partnering with suppliers who will ship products directly to your customers. You can choose from a wide range of niches, such as electronics, home goods, or pet supplies, without having to worry about inventory management. Personally, I've found the most success in ecommerce by using a combination of drop-shipping from certain manufacturers and in-house stocked inventory which my staff ships to our customers.

Using Chat-GPT for market research and finding the best products to sell is discussed later in this book. list of some of the most-searched products that people buy online. Note that this list is based on historical data and trends, and it can change over time as new trends emerge and consumer preferences shift.

1. **Clothing and Apparel:** T-shirts, dresses, shoes, activewear, and fashion accessories are always in high demand online.
2. **Consumer Electronics:** Smartphones, tablets, laptops, cameras, and headphones are popular items for online shoppers.
3. **Beauty and Personal Care Products:** Makeup, skincare, haircare products, and grooming essentials consistently rank high among online purchases.
4. **Home and Kitchen Appliances:** Small appliances like coffee makers, blenders, and air fryers, as well as larger items like refrigerators and washing machines, are commonly bought online.
5. **Toys and Games:** Board games, puzzles, action figures, and educational toys are popular choices for online shoppers.
6. **Health and Wellness Products:** Supplements, vitamins, fitness equipment, and wellness items like essential oils

and diffusers are frequently searched for and purchased online.

7. **Books and eBooks:** Both physical books and digital books continue to be popular among online buyers.
8. **Pet Supplies:** Food, treats, toys, and accessories for pets are commonly bought online, offering convenience and variety for pet owners.
9. **Home Decor and Furniture:** Online shoppers often search for items like wall art, lighting, rugs, and furniture to decorate and furnish their living spaces.
10. **Sports and Outdoor Equipment:** Bicycles, camping gear, exercise equipment, and sporting goods are popular online purchases for outdoor enthusiasts and fitness buffs.

Keep in mind that search trends can vary by region and season, so it's essential to stay up-to-date with the latest trends in your target market to ensure you cater to the most relevant and in-demand products.

More AI Tools for E-Commerce

1. **Puzzle:** AI tool for building a knowledge base for your team and customers.
2. **Civitai:** Civitai is a hub for AI art generation where artists, engineers, and data scientists share and explore models, tutorials, and more.
3. **Cleanup:** AI tool for removing objects, defects, people, or text from pictures.
4. **Browse-AI:** A tool to easily scrape or monitor any website in just 2 minutes with no coding skills.
5. **Brancher-AI:** A no-code platform that lets you build AI-powered apps in minutes with over a hundred ready-made templates to get you started

6. **Do Not Pay:** A platform that offers free legal advice and support to users.
7. **Synthesia-AI**: A platform that will help you make videos quickly and easily.

19

BEST E-COMMERCE SOFTWARE

There are many factors to consider when choosing the right ecommerce software for selling online, such as the number of products, your budget, and the level of custom features you might require. Several ecommerce software brands, however, stand out ahead of the rest and offer an easy way to get started selling online quickly and with little overhead or tech experience. Even better, many of them already employ some AI-based features.

THE FOLLOWING ARE some of the best ecommerce options available, with pros and cons for each.

Shopify

Pros:

- User-friendly interface, suitable for beginners
- Wide variety of templates and themes for store customization

- Large app marketplace for added functionality
- 24/7 customer support

Cons:

- Can become expensive with added apps and premium themes
- Transaction fees apply unless using Shopify Payments
- Limited customization capabilities for advanced users

WooCommerce (for WordPress)

Pros:

- Open-source and highly customizable
- Large plugin marketplace for added functionality
- Integrates seamlessly with WordPress websites
- No platform-specific transaction fees

Cons:

- Steeper learning curve for beginners
- Requires separate hosting and domain
- Additional costs for premium plugins and themes

BigCommerce

Pros:

- Robust feature set built-in, reducing the need for additional apps
- Scalable for growing businesses
- No transaction fees
- Wide variety of themes and customization options

Cons:

- Can become expensive with higher-tier plans
- Limited app marketplace compared to Shopify
- Some features have a learning curve for beginners

Wix eCommerce

Pros:

- Intuitive drag-and-drop website builder
- Wide range of templates and customization options
- Affordable pricing plans
- Integrated marketing and SEO tools

Cons:

- Limited features for larger or more complex stores
- Apps and additional features can be more limited compared to Shopify or WooCommerce
- Migrating to another platform can be challenging

Squarespace

Pros:

- Beautiful, responsive templates designed for various industries
- User-friendly website builder
- Integrated blogging and marketing tools
- Affordable pricing plans

Cons:

- Limited ecommerce features compared to other dedicated platforms
- Smaller app marketplace
- Limited payment gateway options

Personally, I recommend Shopify to anyone who has a desire to start an online store, and I use it for several of my ecommerce stores. It may not be right for everyone, though; again, when choosing ecommerce software for your online store, consider factors such as your level of technical expertise, your store's size and complexity, your budget, and your desired level of customization. It's essential to research each platform and evaluate how well it aligns with your specific needs and goals.

Here are some examples of ways in which AI can be integrated with Shopify ecommerce software, many of which can be applied to other software types as well:

1. **AI Chatbots:** You can integrate AI chatbots like ChatGPT into your Shopify store to handle customer inquiries, provide product recommendations, and assist with the checkout process, reducing the workload on your customer support team.
2. **Inventory Management:** AI-powered inventory management tools can analyze sales data and predict future demand to optimize stock levels, reducing the risk of overstocking or stockouts.
3. **Product Recommendations:** AI can analyze customer browsing and purchase behavior to offer personalized product recommendations, increasing the likelihood of sales and improving the overall shopping experience.

4. **Pricing Optimization:** AI-powered pricing tools can analyze market trends, competitor pricing, and customer behavior to dynamically adjust product prices, ensuring that your store remains competitive while maximizing profit margins.
5. **Email Marketing:** AI can help optimize your email marketing campaigns by analyzing user data to segment your audience, identify the most effective subject lines and content, and determine the optimal sending times.
6. **Fraud Detection:** AI can be integrated with Shopify's fraud detection systems to analyze transactions and detect patterns that indicate potential fraudulent activities, helping to protect your store and customers.
7. **Visual Search:** AI-powered visual search tools allow customers to search for products using images rather than text, making it easier for them to find exactly what they're looking for.
8. **Customer Sentiment Analysis**: AI can analyze customer reviews and feedback to identify trends and sentiment, helping you improve your products and services and address any customer concerns.
9. **Language Translation:** AI-powered translation tools can automatically translate your Shopify store's content into multiple languages, making it accessible to a wider international audience.
10. **Search Engine Optimization (SEO):** AI can help optimize your Shopify store's content and metadata to improve its visibility on search engines, driving more organic traffic to your site.

By integrating AI with Shopify ecommerce software, you can streamline various aspects of your online store, improve the customer experience, and ultimately boost your sales and revenue.

PART III

SUPERCHARGE YOUR BUSINESS

20

ELEVATE YOUR BUSINESS OR BRAND

Unlock the potential of AI to generate innovative ideas and drive business growth

In today's competitive market, businesses and brands must constantly evolve to stay ahead of the curve. Generating new ideas is essential for maintaining relevance, expanding your customer base, and driving growth.

After the success I've had using Chat-GPT to dramatically improve my own online businesses, I'm convinced that there's never been a better time for fellow entrepreneurs to revitalize their existing businesses or brands, all with the expert assistance of Artificial Intelligence. CHAT-GPT, the advanced AI language model that I've come to love after overcoming my concerns about it taking over the world (Spoiler: it won't) offers a valuable resource for businesses and brands looking to generate fresh ideas and revitalize their existing offerings. Let's explore specific ways in which CHAT-GPT can help you breathe new life into your business or brand.

Expanding Product or Service Offerings

Diversifying your product or service offerings can attract new customers and increase revenue. By providing CHAT-GPT with information about your current offerings and target audience, the AI can generate suggestions for complementary products or services that can enhance your existing portfolio.

Prompt: "My business specializes in organic skincare products. Can you suggest new product ideas that would complement my current offerings and appeal to my target audience?"

Improving Customer Experience

Enhancing the customer experience is crucial for boosting satisfaction and loyalty. CHAT-GPT can help you identify areas of improvement by analyzing customer feedback, reviews, and social media conversations. The AI can then provide suggestions for optimizing the customer journey, from user-friendly website design to personalized marketing campaigns.

Prompt: "Based on customer feedback and reviews, how can I improve the overall customer experience for my online clothing store?"

Identifying New Marketing Channels

Reaching your target audience through the right marketing channels is essential for driving sales and brand awareness. CHAT-GPT can help you explore untapped marketing channels and assess their potential effectiveness for your business, such as influencer marketing, podcast advertising, or content marketing.

Prompt: "My brand specializes in sustainable fashion. Can you suggest new marketing channels that would help me reach a wider audience and increase brand visibility?"

Refining Brand Messaging

A clear and consistent brand message is key to connecting with your target audience. CHAT-GPT can help you refine your brand messaging by analyzing your current communication efforts and suggesting improvements, such as emphasizing your unique selling points, adopting a more conversational tone, or incorporating storytelling.

Prompt: "My brand focuses on artisanal, handcrafted home décor. How can I refine my brand messaging to better convey our values and resonate with our target audience?"

Exploring New Target Markets

Expanding into new target markets can significantly increase your customer base and revenue. CHAT-GPT can help you identify potential target markets that align with your business values and offerings. The AI can also provide insights into the preferences and pain points of these new markets, enabling you to tailor your marketing strategy accordingly.

Prompt: "My business offers premium, small-batch coffee beans. Can you suggest new target markets that might be interested in our products, and provide insights into their preferences and needs?"

Innovating with Technology

Adopting new technology can streamline operations, enhance customer experience, and differentiate your business from competitors. CHAT-GPT can help you identify innovative technologies relevant to your industry, such as virtual reality, IoT, or AI-powered analytics, and explore potential use cases for your business.

Prompt: "I run a fitness studio that offers group classes and personal training. Can you suggest innovative technologies that could improve our services and customer experience?"

Developing Strategic Partnerships

Forming strategic partnerships can open up new opportunities for growth and expansion. CHAT-GPT can help you identify potential partners, such as complementary businesses or industry influencers, who share your vision and can contribute to your business goals. The AI can even help you draft pitches to reach out to these potential partners.

Prompt: "My eco-friendly cleaning products company is looking to form strategic partnerships. Can you suggest potential partners that align with our values and provide ideas for collaboration opportunities?"

Whether you're looking to expand your product offerings, refine your brand messaging, or explore new markets, CHAT-GPT is a powerful tool for generating innovative ideas that can revitalize your existing business or brand. By utilizing the capabilities of this advanced AI language model, you can unlock new growth opportunities, adapt to changing market conditions, and maintain a competitive edge.

21

REVITALIZE YOUR BUSINESS PLAN

Unlock the full potential of AI to refine your business strategies and achieve growth

Effective business planning is crucial for the growth and success of any business, regardless of its size or industry. Chat-GPT, an advanced AI language model, offers valuable insights and guidance for refining your existing business plan and adapting it to the ever-changing market landscape. Let's explore specific ways in which Chat-GPT can help you enhance your business planning, along with example prompts to guide your interaction with this powerful AI tool.

Revisiting Your Executive Summary

An up-to-date executive summary is essential for keeping stakeholders informed about your business's progress and direction. Chat-GPT can help you review and refine your existing summary to reflect your current objectives and achievements.

Example Prompt: "My business has pivoted from a subscription box service to a marketplace for eco-friendly products. Can you help me rewrite our executive summary to reflect this change in direction?"

Market Analysis and Competitive Landscape

Market conditions and competition evolve over time, making it crucial to periodically reassess your market analysis. Chat-GPT can assist you in evaluating the current market situation and identifying emerging trends and competitors.

Example Prompt: "My brand focuses on organic skincare products. Can you help me analyze the current market trends and identify new competitors in this space?"

Marketing and Sales Strategy Review

As consumer behavior and preferences change, it's essential to adapt your marketing and sales strategies accordingly. Chat-GPT can help you identify new advertising channels, promotional tactics, and pricing models to boost sales and engagement.

Example Prompt: "My business offers online language courses. Can you help me review our marketing and sales strategies and suggest improvements to reach a broader audience and increase conversions?"

Streamlining Operations and Supply Chain

Efficient operations and supply chain management are vital for maintaining profitability and customer satisfaction. Chat-GPT can provide insights and recommendations for optimizing your processes, logistics, and inventory management.

Example Prompt: "My business imports and sells artisanal coffee beans. Can you help me identify potential bottlenecks in our supply chain and suggest ways to streamline our operations?"

Financial Analysis and Projections

Regular financial analysis and forecasting are crucial for tracking your business's performance and planning for future growth. Chat-GPT can help you review your financial data, identify areas of improvement, and develop updated financial projections.

Example Prompt: "My e-commerce store sells custom-printed merchandise. Can you help me analyze our financial performance over the past year and create updated financial projections for the next three years?"

Assessing and Expanding Product Offerings

Adapting your product offerings to market demand can be vital for your business's success. Chat-GPT can help you evaluate your current product line and explore new opportunities for product development and expansion.

Example Prompt: "My company specializes in sustainable clothing. Can you help me assess our current product offerings and suggest new product ideas that align with our brand values and market demand?"

Employee Training and Development

Investing in employee training and development is essential for fostering a skilled and motivated workforce. Chat-GPT can help you design training programs and identify areas for skill development and growth.

Example Prompt: "My business operates a chain of boutique hotels. Can you help me create an employee training program to improve customer service and enhance guest experiences?"

Harnessing the power of Chat-GPT can be a game-changer for your existing business or brand. By leveraging the insights and guidance provided by this advanced AI language model, you can refine your business planning, optimize your strategies, and set your business on a path toward growth and success.

22

ENERGIZE YOUR CONTENT CREATION

Enhance your brand's storytelling and engage your audience with AI-powered content

In today's fast-paced digital landscape, high-quality content creation and copywriting are more important than ever for businesses to stand out from the competition. However, producing engaging content consistently can be challenging. Chat-GPT, an AI-powered language model, can help businesses overcome these challenges by generating compelling content tailored to their brand voice and audience. Let's explore specific ways to use Chat-GPT for content creation and copywriting, along with example prompts to help you get started.

Revamping Website Content

A well-crafted website is crucial for showcasing your brand's identity and values. Chat-GPT can help you refresh your website content, making it more engaging and relevant to your target audience.

Example Prompt: "My fashion brand has recently expanded to include sustainable accessories. Can you help me rewrite our 'About Us' page to reflect our new eco-friendly focus?"

Content Strategy and Editorial Calendar

A comprehensive content strategy is essential for achieving your marketing goals. Chat-GPT can help you brainstorm content ideas and develop an editorial calendar that aligns with your business objectives and audience interests.

Example Prompt: "My online marketing agency specializes in SEO and PPC. Can you help me create a 3-month editorial calendar with blog post topics that highlight our expertise?"

Refreshing Old Blog Posts

Updating your existing blog posts with new information and insights can improve their search engine ranking and keep your content relevant. Chat-GPT can help you revise and expand old blog posts to ensure they continue to provide value to your audience.

Example Prompt: "I have a blog post about the top 5 social media trends from 2021. Can you help me update it to reflect the latest trends and best practices in 2023?"

Crafting Social Media Updates

Consistently creating captivating social media content can be a time-consuming task. Chat-GPT can help you develop engaging copy for various platforms, maintaining your brand voice and staying relevant to your audience.

Example Prompt: "My bakery specializes in custom-made cakes. Can you help me write a series of Facebook posts showcasing our cake designs and promoting our cake consultation services?"

Email Marketing Content

Personalized and engaging email campaigns can boost customer retention and drive conversions. Chat-GPT can help you craft compelling email content, including subject lines, promotional offers, and newsletters.

Example Prompt: "My online fitness platform is launching a new workout program. Can you help me write a series of emails to inform our subscribers about the program and encourage them to sign up?"

Case Studies and Customer Testimonials

Showcasing your business's success stories and customer experiences can build trust and credibility. Chat-GPT can help you write persuasive case studies and customer testimonials that highlight the value of your products or services.

Example Prompt: "My software company offers a CRM solution for small businesses. Can you help me write a case study that demonstrates how our CRM has helped a small business improve their sales and customer relationships?"

Whitepapers and E-books

Publishing in-depth whitepapers and e-books can help you position your brand as an industry expert. Chat-GPT can assist you in creating well-researched, informative, and engaging long-form content that appeals to your target audience.

Example Prompt: "My financial consulting firm specializes in tax planning. Can you help me write a whitepaper on the latest tax strategies for small businesses to save money and optimize their finances?"

Leveraging the power of Chat-GPT for content creation and copywriting can be a game-changer for your existing business or brand. With AI-generated content tailored to your brand voice and audience, you can deliver high-quality, engaging content consistently across various platforms. By using Chat-GPT to elevate your content strategy, you can connect more effectively with your target audience, build trust and credibility, and drive business growth. Embracing the potential of AI to streamline your content creation process will allow you to focus on other critical aspects of your business and set your brand apart from the competition.

23

AMPLIFY YOUR SOCIAL MEDIA

Enhance your social media presence and drive engagement with AI-powered content and insights

Social media management is crucial for businesses looking to stay relevant, engage with their audience, and drive growth. With the help of Chat-GPT, an AI-powered language model, businesses can streamline their social media management efforts and generate compelling content tailored to their brand voice and target audience. Let's explore specific ways to use Chat-GPT for social media management for your existing business or brand, along with example prompts to help you get started.

Social Media Platform Expansion

Expanding your presence to additional social media platforms can help you reach new audiences and further engage with your existing customers. Chat-GPT can help you identify which platforms align with your brand and target audience, as well as suggest strategies for success.

Example Prompt: "My eco-friendly home goods brand is

currently on Instagram and Facebook. Can you help me determine if expanding to Pinterest or TikTok would be beneficial and suggest strategies for those platforms?"

Social Media Content Refresh

Keeping your social media content fresh and relevant is essential for maintaining audience engagement. Chat-GPT can help you revitalize your content by generating new ideas or suggesting ways to repurpose existing content.

Example Prompt: "My fitness studio has been posting workout tips and motivational quotes on Instagram. Can you help me come up with new content ideas or ways to repurpose our existing content for more engagement?"

Social Media Content Calendar

A well-organized content calendar can help you maintain a consistent posting schedule and ensure a balanced content mix. Chat-GPT can help you plan your social media content calendar, outlining the types of content and the optimal posting frequency.

Example Prompt: "My marketing agency specializes in SEO and content marketing. Can you help me create a two-month social media content calendar for LinkedIn that focuses on showcasing our expertise and driving leads?"

Competitor Analysis and Benchmarking

Understanding your competitors' social media strategies can help you identify opportunities for growth and improvement. Chat-GPT can help you analyze your competitors' social media presence and provide insights for refining your own approach.

Example Prompt: "My online clothing store has three main competitors. Can you help me analyze their social media presence and suggest ways we can differentiate ourselves and improve our strategy?"

Social Media Advertising Copy

Effective social media advertising can help you reach a broader audience and drive conversions. Chat-GPT can help you craft persuasive ad copy tailored to your target audience and optimized for each platform's specifications.

Example Prompt: "My e-commerce business sells customizable phone cases. Can you help me write ad copy for a Facebook and Instagram ad campaign that highlights our unique designs and customization options?"

Social Media Contests and Giveaways

Contests and giveaways can boost audience engagement, increase brand awareness, and encourage user-generated content. Chat-GPT can help you plan and execute successful social media contests or giveaways that align with your brand goals.

Example Prompt: "My skincare brand is looking to increase our Instagram followers and engagement. Can you help me come up with a creative contest or giveaway idea that encourages users to participate and share our content?"

Social Media Crisis Management

Handling social media crises or negative feedback effectively is vital for protecting your brand's reputation. Chat-GPT can help you draft thoughtful responses to address concerns, clarify misunderstandings, or provide information.

Example Prompt: "A customer left a negative review on our Facebook page about a delayed shipment. Can you help me write a response that addresses their concerns, apologizes for the inconvenience, and provides a solution?"

Leveraging Chat-GPT for social media management can significantly enhance your existing business or brand's online presence. By utilizing AI-generated content and insights, you can streamline your social media efforts, maintain a consistent brand voice, and effectively engage with your target audience. Using AI to optimize your social media strategy will let you focus on other critical aspects of your business and set your brand apart from the competition.

More AI Tools for Growing Your Business

Playground-AI: online platform that allows you to edit your photos using just text.

Poised: A website that provides information and resources on personal development and self-improvement.

Detangle-AI: A tool that uses AI to simplify complex legal documents into easy-to-understand summaries.

Runway: Features over 30 "Magic Tools" for editing and enhancing your content.

Eye Contact By NVIDIA: A tool for anyone looking to enhance their video and audio during livestreams, video chats, or video conference calls.

24

SUPERCHARGE YOUR SEO

Harness the power of AI to improve your online store's SEO and attract more customers

In the competitive world of e-commerce, search engine optimization (SEO) is crucial for driving organic traffic and increasing sales. As a powerful AI language model, Chat-GPT can be an invaluable tool in optimizing your online store for search engines. Let's explore how you can use Chat-GPT to improve your online store's SEO and provide examples of prompts to help you make the most of this advanced technology.

**Also see our previous article in the E-Commerce section of this book for detailed information about the best e-commerce software and how to use Shopify with AI.*

Keyword Research

Identifying the right keywords is essential for optimizing your online store's content. Chat-GPT can help you generate keyword ideas related to your products or industry.

Example Prompt: "Generate a list of 20 long-tail keywords related to sustainable clothing."

Optimizing Product Descriptions

Well-written and SEO-friendly product descriptions can improve your store's visibility on search engines. Chat-GPT can help you craft engaging, keyword-rich descriptions for your products.

Example Prompt: "Write a 150-word SEO-optimized product description for a reusable coffee cup made from bamboo."

Creating Compelling Meta Descriptions and Title Tags

Meta descriptions and title tags are critical components of on-page SEO. Chat-GPT can help you generate persuasive and keyword-optimized meta descriptions and title tags for your online store's pages.

Example Prompt: "Create an SEO-optimized title tag and meta description for a webpage selling eco-friendly yoga mats."

Building SEO-Friendly Blog Content

Regularly publishing high-quality, SEO-friendly blog content can help you attract organic traffic and establish your online store as an authority in your niche. Chat-GPT can help you generate blog post ideas and draft engaging, keyword-rich content.

Example Prompt: "Write an outline for a 1,000-word blog post about the benefits of organic skincare products."

Crafting Engaging Social Media Content

Sharing valuable, SEO-friendly content on social media can help you drive traffic to your online store. Chat-GPT can help you create captivating social media posts that incorporate targeted keywords and hashtags.

Example Prompt: "Compose a tweet promoting a new line of vegan leather bags, including relevant keywords and hashtags."

Analyzing Competitors

Understanding your competitors' SEO strategies can help you identify opportunities for improvement. Chat-GPT can help you generate a list of questions to ask when analyzing your competitors' SEO efforts.

Example Prompt: "List 10 questions to ask when analyzing a competitor's SEO strategy in the handmade jewelry market."

Developing a Comprehensive SEO Strategy

To achieve the best results, your online store needs a comprehensive SEO strategy. Chat-GPT can help you outline the key components of an effective SEO plan for your e-commerce business.

Example Prompt: "Outline the main steps in creating a successful SEO strategy for an online store specializing in gourmet chocolates."

When you harness the power of Chat-GPT, you can significantly improve your online store's search engine optimization and attract more customers. From keyword research and content creation to competitor analysis and strategic planning, Chat-GPT can help you tackle a wide array of SEO tasks.

25

MODERNIZE YOUR CUSTOMER SUPPORT

Improve customer satisfaction and streamline support processes with AI-powered assistance

Providing exceptional customer support is essential for any business looking to retain customers and build a loyal customer base. Chat-GPT, an AI-powered language model, can help businesses improve their customer support processes by generating tailored responses, assisting in problem-solving, and addressing customer concerns more efficiently. Let's explore specific ways to use Chat-GPT for customer support for your existing business or brand, along with example prompts to help you get started.

**Also see our previous article in the Start-ups section of this book for detailed information on customer support software and how to use Zendesk with AI.*

FAQ and Help Center Content

Creating comprehensive and easy-to-understand help center content can reduce the number of customer inquiries and improve overall customer satisfaction. Chat-GPT can help you develop or update your FAQ and help center articles to better address customer questions.

Example Prompt: "My e-commerce website sells electronic gadgets. Can you help me create an FAQ section that addresses common customer questions about shipping, returns, and warranty information?"

Email Support Template Creation

Having a set of well-crafted email templates can save your customer support team time and ensure consistent responses. Chat-GPT can help you develop email templates for various support scenarios that maintain your brand voice and address customer concerns effectively.

Example Prompt: "My online course platform offers a variety of courses and payment plans. Can you help me create an email template for responding to customers who inquire about refunds or cancellations?"

Live Chat Support Assistance

Chat-GPT can assist your live chat support team by suggesting responses to customer inquiries. By using AI-generated suggestions, your team can respond to customer questions more efficiently, improving response times and customer satisfaction.

Example Prompt: "A customer in our live chat is asking about the compatibility of our software with their specific operating system. Can you help me draft a response that explains the system requirements and provides a link to our compatibility guide?"

Social Media Customer Support

Monitoring and responding to customer inquiries and feedback on social media platforms is crucial for maintaining a positive brand image. Chat-GPT can help you draft personalized and thoughtful responses to customer messages or comments on your social media channels.

Example Prompt: "A customer commented on our Instagram post asking about the sizing of our clothing products. Can you help me write a response that directs them to our sizing guide and offers assistance if they have further questions?"

Knowledge Base Development

A well-organized knowledge base can empower your customers to find solutions to their issues independently, reducing the workload on your support team. Chat-GPT can help you create or update your knowledge base articles to ensure they are clear, informative, and easy to follow.

Example Prompt: "My project management software has recently added new features. Can you help me write a knowledge base article that explains how to use the new time tracking functionality?"

Support Chatbot Training

Chatbots can handle a high volume of customer inquiries, but they require a robust knowledge base and training to provide accurate and helpful responses. Chat-GPT can help you train your chatbot by generating realistic customer inquiries and appropriate responses based on your business's specific needs.

Example Prompt: "My travel agency offers vacation packages and deals. Can you help me generate a list of 20 realistic customer inquiries and appropriate responses for training our customer support chatbot?"

Customer Feedback Analysis

Analyzing customer feedback is crucial for identifying areas of improvement and understanding customer needs. Chat-GPT can help you review customer feedback and provide insights and recommendations for enhancing your support processes.

Example Prompt: "We have received mixed feedback about our customer support response times. Can you help me analyze the feedback and suggest ways to improve our response times and customer satisfaction?"

Integrating Chat-GPT into your customer support processes can significantly improve the customer experience and streamline your support efforts. By leveraging AI-generated content and insights, you can ensure consistent, timely, and personalized responses to customer inquiries, ultimately leading to increased customer satisfaction and loyalty. Now is the time to start getting ahead of the curve and driving business growth with the help of Chat-GPT and its innovative capabilities in the realm of customer support.

26

CREATE YOUR A.I. SUPPORT AGENT

To use Chat-GPT as a customer support chatbot for your ecommerce website, you will need to follow these steps:

1. **Sign up for an API key:** You will need to sign up for an API key with OpenAI to gain access to the Chat-GPT API. You can do this by visiting the OpenAI website (https://www.openai.com/) and signing up for a suitable pricing plan.
2. **Develop the chatbot interface:** Create a user interface for your chatbot, which can be integrated into your ecommerce website. This interface should include a chat window where users can type their queries and receive responses from the chatbot.
3. **Connect to the Chat-GPT API:** You will need to connect your chatbot interface to the Chat-GPT API using the API key provided by OpenAI. To do this, you will need some programming knowledge or the assistance of a developer. OpenAI provides documentation on how to make API calls and handle responses: https://platform.openai.com/docs/api-reference/introduction

4. **Structure the conversation:** When sending user queries to the Chat-GPT API, you may want to provide some context or conversation history to get more accurate responses. The API allows you to send a series of messages as input, including prior customer messages and chatbot responses. This helps the model maintain context and provide better answers.
5. **Set safety and content filters:** To ensure that the responses generated by Chat-GPT align with your brand values and avoid inappropriate content, you can implement safety and content filters. You can refer to OpenAI's moderation guide for more information: https://platform.openai.com/docs/guides/moderation
6. **Test and iterate:** Before deploying the chatbot on your website, test it thoroughly by simulating various customer queries and use cases. This will help you identify any issues or areas where the chatbot can be improved.
7. **Integrate and deploy:** Once you are satisfied with the chatbot's performance, integrate it into your ecommerce website and make it available to your customers. Monitor the chatbot's performance and gather user feedback to make ongoing improvements.

Remember that while Chat-GPT can be an excellent customer support tool, it may still require supervision and fine-tuning to deliver the best possible results. Consider having a human customer support team in place to handle complex queries or situations where the chatbot might not provide adequate assistance.

27

SUPERCHARGE YOUR MARKET RESEARCH

Discover how AI can provide valuable insights into your industry, competition, and target audience

Market research is not only crucial for starting a business but also for growing and maintaining an existing one. With Chat-GPT, a powerful AI language model, businesses can uncover valuable insights, stay ahead of trends, and make data-driven decisions. Let's explore specific ways to use Chat-GPT for market research for your existing business or brand, along with example prompts to help you get started.

Monitoring Industry Trends and Innovations

Keeping up with industry trends and innovations is essential for staying competitive and identifying new opportunities. Chat-GPT can help you stay informed about the latest developments in your industry and assess their impact on your business.

Example Prompt: "My business is in the renewable energy sector. Can you help me identify and analyze recent innovations and trends in renewable energy and their potential impact on my company?"

Assessing Competitors' Strategies

Understanding your competitors' strategies can provide valuable insights into your market positioning and help you refine your own approach. Chat-GPT can help you analyze your competitors and their tactics, giving you an edge in your industry.

Example Prompt: "My company sells premium skincare products. Can you help me analyze the marketing strategies of my top three competitors and suggest ways we can improve our own strategy?"

Expanding Your Product or Service Offering

When considering expanding your product or service offering, it's essential to conduct market research to identify gaps and opportunities. Chat-GPT can assist you in evaluating customer needs and trends to inform your expansion plans.

Example Prompt: "My brand offers a line of eco-friendly cleaning products. Can you help me identify potential gaps in the market and opportunities for expanding our product range?"

Refining Your Target Audience

As your business grows, it's crucial to refine and adjust your target audience to better serve your customers. Chat-GPT can help you update your buyer personas and identify new ways to reach and engage with your audience.

Example Prompt: "My online fitness coaching business has grown, and I want to refine my target audience. Can you help me update my buyer personas and suggest strategies for reaching and engaging with this audience?"

Identifying New Market Opportunities

Expanding your business into new markets can be a lucrative opportunity, but it requires thorough research. Chat-GPT can help you evaluate potential new markets, assess competition, and identify entry barriers to inform your expansion strategy.

Example Prompt: "My subscription box service for pet products has been successful in the US market. Can you help me evaluate the potential for expanding into the European market and identify key competitors and challenges?"

Analyzing Customer Feedback

Customer feedback is a valuable source of information for improving your products or services and identifying areas for growth. Chat-GPT can help you analyze customer feedback and provide insights and recommendations for enhancements.

Example Prompt: "My e-commerce platform has received mixed reviews about our user interface. Can you help me analyze the feedback and suggest improvements to enhance the user experience?"

Evaluating Pricing Strategies

As your business evolves, it's important to periodically assess your pricing strategy to ensure it remains competitive and profitable. Chat-GPT can help you analyze pricing strategies of competitors, market conditions, and customer expectations to develop an updated pricing model.

Example Prompt: "My software as a service (SaaS) company has been growing, and I want to evaluate our current pricing strategy. Can you help me analyze the pricing models of similar SaaS companies and suggest adjustments to balance competitiveness and profitability?"

Chat-GPT is a powerful tool for market research. Never before in history has there been such ease of access to valuable insights that can help your existing business or brand stay ahead of the competition.

28

CREATE YOUR A.I. MARKET RESEARCHER

Using Chat-GPT for market research for an ecommerce business can provide valuable insights to help you understand your target audience, competition, and trends.

Here's a step-by-step guide based on one of my ecommerce stores, *CandidCandle.com*, which sells handmade scented candles:

1. **Access Chat-GPT:** Sign up for an API key with OpenAI (https://www.openai.com/) to gain access to the Chat-GPT API. Choose a suitable pricing plan and familiarize yourself with the API documentation (https://platform.openai.com/docs/api-reference/introduction).
2. **Define your market research objectives:** Before using Chat-GPT for market research, identify your objectives. These might include understanding customer preferences, identifying trends, evaluating competitors, or discovering new product ideas.
3. **Create specific prompts:** Develop clear and concise prompts that will help you gather the information you need. For example:

4. *"What are the most popular scented candle fragrances in 2023?"*
5. *"What are the current trends in scented candles?"*
6. *"Who are the top competitors in the scented candles market?"*
7. *"What are the key factors customers consider when buying scented candles online?"*
8. **Send prompts to the Chat-GPT API:** Using the API key, connect to the Chat-GPT API and send the prompts you created. You might need programming knowledge or assistance from a developer for this step. Remember to structure your queries effectively, providing context and being specific about the information you need.
9. **Analyze and interpret the responses:** Chat-GPT will generate responses based on the prompts you send. Carefully analyze these responses to extract useful information and insights. Keep in mind that the information provided by Chat-GPT may not always be up-to-date, so it's essential to verify the data through additional research.
10. **Combine with other research methods:** While Chat-GPT can provide valuable information, it's essential to complement its insights with other market research methods, such as surveys, interviews, and analysis of social media platforms and forums where your target audience discusses scented candles.
11. **Apply findings to your business strategy:** Use the insights you've gained from Chat-GPT and other research methods to refine your marketing strategies, product offerings, and overall business plan. This could involve updating your product line, adjusting your pricing strategy, or improving your marketing efforts.
12. **Continuously improve:** Market research is an ongoing process. Periodically revisit your research objectives and prompts to stay up-to-date with market trends and customer preferences. Use Chat-GPT to support your

ongoing research efforts and make data-driven decisions for your scented candles ecommerce business.

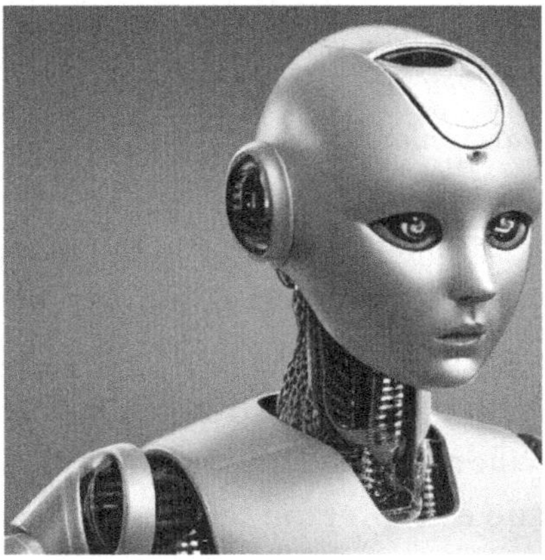

29

DEVELOP INCREDIBLE PRODUCTS

Unlock the power of AI to enhance your product offerings, designs, and market positioning

For businesses looking to grow, product development is an ongoing process that can lead to increased market share, customer satisfaction, and innovation. Chat-GPT, an advanced AI language model, offers businesses a valuable resource for generating ideas, refining designs, and optimizing products. Let's explore specific ways to use Chat-GPT for product development in your existing business or brand, along with example prompts to help you get started.

Expanding Your Product Line

Chat-GPT can help you brainstorm new product ideas that align with your brand and cater to your target audience, expanding your product offerings and driving growth.

Example Prompt: "My business offers eco-friendly cleaning products. Can you help me generate ideas for additional products that complement our existing line and cater to our environmentally conscious customers?"

Enhancing Existing Products

Continuously improving your existing products can lead to increased customer satisfaction and loyalty. Chat-GPT can provide suggestions for enhancements based on user feedback, industry trends, and best practices.

Example Prompt: "Our business sells wireless earbuds, and we'd like to improve their battery life and overall performance. Can you suggest potential enhancements to our current design and features?"

Identifying Product Line Extensions

Extending your product line can help you reach new customer segments and increase market share. Chat-GPT can help you identify potential product extensions that resonate with your target audience.

Example Prompt: "My brand sells natural skincare products. Can you help me identify potential product line extensions that cater to different skin types and concerns?"

Optimizing Product Packaging and Design

The packaging and design of your products play a crucial role in attracting customers and communicating your brand's values. Chat-GPT can provide design suggestions and improvements based on industry best practices and user preferences.

Example Prompt: "Our business offers organic snacks, and we'd like to revamp our packaging to better convey our brand's commitment to sustainability. Can you suggest design improvements for our packaging that align with this message?"

Soliciting and Analyzing Customer Feedback

Customer feedback is essential for understanding how your products can be improved and identifying areas for growth. Chat-GPT can help you analyze customer feedback and provide insights and recommendations for enhancements.

Example Prompt: "My e-commerce platform has received mixed reviews about our mobile app. Can you help me analyze the feedback and suggest improvements to enhance the user experience?"

Evaluating and Updating Pricing Strategies

Periodically assessing your pricing strategy is crucial to ensure it remains competitive and profitable. Chat-GPT can help you analyze pricing strategies of competitors, market conditions, and customer expectations to develop an updated pricing model.

Example Prompt: "My software as a service (SaaS) company has been growing, and I want to evaluate our current pricing strategy. Can you help me analyze the pricing models of similar SaaS companies and suggest adjustments to balance competitiveness and profitability?"

Assessing the Impact of New Technologies

Staying current with technological advancements is essential for maintaining a competitive edge. Chat-GPT can help you identify emerging technologies that could impact your products and provide insights on how to adapt and integrate them.

Example Prompt: "My company manufactures smartwatches. Can you help me identify recent technological advancements in wearable technology and suggest ways we can incorporate them into our products?"

When using Chat-GPT for help with product development,

existing businesses and brands can unlock new opportunities for growth, innovation, and differentiation.

30

CREATE YOUR A.I. PRODUCT DEVELOPER

Using Chat-GPT for product development can help you generate creative ideas and make data-driven decisions for your business. Here's a step-by-step guide based on my scented candle business, Candid Candle:

1. **Access Chat-GPT:** Sign up for an API key with OpenAI (https://www.openai.com/) to gain access to the Chat-GPT API. Choose a suitable pricing plan and familiarize yourself with the API documentation (https://platform.openai.com/docs/api-reference/introduction).
2. **Define your product development objectives:** Identify your goals for product development. These might include creating new scents, improving packaging, finding sustainable materials, or expanding your product line.
3. **Create specific prompts:** Develop clear and concise prompts that will help you gather the information and ideas you need. For example, I used the following prompts to expand my product line on CandidCandle.com:

4. *"What are some unique scented candle fragrance ideas?"*
5. *"How can I make my scented candles more eco-friendly?"*
6. *"What are some innovative packaging ideas for scented candles?"*
7. *"What are some successful product line extensions for scented candle businesses?"*
8. **Send prompts to the Chat-GPT API:** Using the API key, connect to the Chat-GPT API and send the prompts you created. You might need programming knowledge or assistance from a developer for this step. Remember to structure your queries effectively, providing context and being specific about the information you need.
9. **Analyze and interpret the responses:** Chat-GPT will generate responses based on the prompts you send. Carefully analyze these responses to extract useful ideas and insights. Remember to evaluate the feasibility and market potential of these ideas before implementing them.
10. **Gather feedback and test ideas:** Engage with your customers and target audience to gather feedback on the ideas generated by Chat-GPT. Conduct surveys, focus groups, or social media polls to understand how your audience perceives these new ideas. This will help you refine your product development process.
11. **Prototype and iterate:** Based on the feedback and insights, create prototypes or samples of the new products. Test these prototypes with customers or a small focus group, and iterate based on their feedback to improve the final product.
12. **Launch and promote your new products:** Once you've refined your new scented candles, launch them and promote them through your marketing channels, such as your website, social media, email campaigns, and more.

13. **Continuously improve:** Product development is an ongoing process. Periodically revisit your product development objectives and prompts to stay up-to-date with market trends, customer preferences, and emerging technologies. Use Chat-GPT to support your ongoing product development efforts and make data-driven decisions for your scented candles business.

31

BREATHTAKING BRANDING

Leverage the power of AI to amplify your personal brand and build impactful influencer collaborations

In today's competitive market, branding and influencer marketing are crucial for the growth and success of existing businesses and brands. Chat-GPT, an advanced AI language model, offers an array of applications to help entrepreneurs enhance their personal brand and optimize influencer marketing campaigns. Let's explore specific ways to use Chat-GPT for personal branding and influencer marketing for your existing business or brand, along with example prompts to help you get started.

Refining Your Brand

Your brand is an evolving representation of your values, expertise, and unique selling points. Chat-GPT can help you reassess and refine your personal brand, ensuring that it remains relevant and resonates with your target audience.

Example Prompt: "My brand focuses on digital marketing. Can you help me refine my brand to better align with the current trends and appeal to my target audience?"

Enhancing Your Brand Story

A captivating brand story can deepen your connection with your audience. Chat-GPT can assist you in revisiting and enhancing your narrative, ensuring it effectively communicates your journey, expertise, and values.

Example Prompt: "I'm a nutritionist who helps clients improve their eating habits. Can you help me enhance my brand story to better showcase my passion for nutrition and the results my clients have achieved?"

Generating Fresh Content Ideas

Keeping your content fresh and engaging is key to maintaining your audience's interest. Chat-GPT can generate new content ideas and draft blog posts, social media updates, or video scripts that align with your brand and captivate your audience.

Example Prompt: "I'm a leadership coach, and I need fresh content ideas for my YouTube channel. Can you suggest engaging topics that will resonate with my target audience and showcase my expertise?"

Discovering New Influencers for Collaboration

Collaborating with new influencers can expand your reach and keep your marketing efforts fresh. Chat-GPT can help you identify potential influencers within your niche who align with your brand values and goals.

Example Prompt: "My business offers sustainable fashion products. Can you help me identify new influencers in the fashion industry who share similar values and might be interested in collaborating with my brand?"

Crafting Tailored Influencer Outreach Messages

To forge successful influencer collaborations, personalized and compelling outreach messages are essential. Chat-GPT can help you draft targeted outreach messages that showcase your brand and highlight potential collaboration opportunities.

Example Prompt: "I'm looking to collaborate with a tech influencer for my software company. Can you help me draft a personalized outreach message that outlines the benefits of partnering with my brand?"

Creating Innovative Influencer Marketing Campaigns

Innovative influencer marketing campaigns can boost brand awareness and drive conversions. Chat-GPT can help you generate unique campaign ideas that leverage the strengths of your chosen influencers and captivate their audience.

Example Prompt: "I've partnered with a fitness influencer for my activewear brand. Can you help me come up with a creative influencer marketing campaign that showcases our products and resonates with their audience?"

Optimizing Influencer Marketing Performance

Measuring and optimizing your influencer marketing efforts is crucial for maximizing your return on investment. Chat-GPT can help you analyze campaign results, identify areas for improvement, and suggest ways to enhance future collaborations.

Example Prompt: "I recently ran an influencer marketing campaign for my subscription box service. Can you help me evaluate the results and recommend strategies to improve our influencer marketing performance moving forward?"

By turning to Chat-GPT for personal branding and influencer marketing, existing businesses and brands can increase their reach in dramatic ways.

32

CAPTIVATE CUSTOMERS

How AI-driven language models can supercharge your company's growth

Growing your customer base and revenue are top priorities for any business. Chat-GPT, a state-of-the-art AI language model, can be a valuable tool in achieving these goals. Let's explore the different ways Chat-GPT can help increase your company's customers and revenue and provide examples of prompts to ensure optimal results.

Crafting Persuasive Sales Copy

Compelling sales copy can effectively communicate the value of your products or services, converting prospects into customers. Chat-GPT can help you create persuasive sales copy for your website, ads, or marketing materials.

Example Prompt: "Write a persuasive product description for our eco-friendly, waterproof backpack, highlighting its durability, spaciousness, and sustainable materials."

Enhancing Email Marketing Campaigns

Email marketing remains an essential channel for customer acquisition and retention. Chat-GPT can help you craft engaging and personalized email content that encourages readers to take action, such as purchasing your products or services.

Example Prompt: "Create a series of three engaging email sequences to onboard new subscribers and promote our upcoming sale on eco-friendly outdoor gear."

Creating Engaging Blog Posts and Articles

Informative and well-written blog posts or articles can attract new visitors to your website and help establish your company as an industry expert. Chat-GPT can help you create content that resonates with your target audience and drives traffic to your website.

Example Prompt: "Write a 1,000-word blog post on the benefits of eco-friendly outdoor gear and the importance of sustainability in the outdoor industry."

Optimizing Social Media Content

Social media platforms provide a powerful way to reach potential customers and strengthen brand loyalty. Chat-GPT can help you create captivating social media content that grabs your audience's attention and drives engagement.

Example Prompt: "Generate ten attention-grabbing Instagram captions for our upcoming eco-friendly outdoor gear product launch."

Developing Effective PPC Ad Campaigns

Pay-per-click (PPC) advertising is a cost-effective method for acquiring new customers. Chat-GPT can help you craft compelling ad copy that drives clicks and conversions.

Example Prompt: "Write three different Google Ads headlines and descriptions for our eco-friendly, waterproof backpack, targeting outdoor enthusiasts who value sustainability."

Crafting Conversion-Boosting Landing Pages

A well-designed landing page can significantly increase conversion rates for your products or services. Chat-GPT can help you create persuasive and relevant landing page copy that guides visitors towards taking the desired action.

Example Prompt: "Write landing page copy for our eco-friendly outdoor gear sale, emphasizing the limited-time offer, product benefits, and our commitment to sustainability."

Streamlining Customer Support

Providing exceptional customer support can enhance customer satisfaction and promote repeat business. Chat-GPT can help you create comprehensive FAQ pages, chatbot scripts, or canned responses for common customer inquiries.

Example Prompt: "Create ten frequently asked questions and answers about our eco-friendly, waterproof backpack, addressing product features, warranty, and shipping."

Using Chat-GPT, your company can generate persuasive and engaging content that attracts new customers and boosts revenue. From crafting compelling sales copy to optimizing your social media presence, this advanced AI language model can be a game-changer for businesses looking to scale and thrive in today's competitive landscape.

More AI Tools for Business

Otter (https://otter.ai/): AI-powered online software for capturing and sharing insights from meetings, making it easier for teams to stay on top of important discussions and decisions.

Krisp (https://krisp.ai/): AI-powered online software that removes background voices, noises, and echoes from calls

Unbounce Smart Copy (https://unbounce.com/): AI tool for writing high-performing cold emails at scale. It helps you write emails that get opened, clicked and replied to, so you can focus on closing deals.

33

REVOLUTIONIZE YOUR VISUAL CONTENT

Harness the power of AI to create stunning visuals and make a lasting impression on your audience

In today's digital landscape, visual content and graphic design are paramount for capturing the attention of your target audience. AI-powered tools, including Chat-GPT, can help existing businesses and brands streamline the process of creating eye-catching visuals and designs. Let's explore specific ways to use AI for visual content and graphic design, along with example prompts to help you get started with Chat-GPT.

Brainstorming Creative Design Concepts

AI tools like Chat-GPT can help you come up with innovative design concepts for your marketing materials, ensuring they stand out from the competition.

Example Prompt: "I need a creative design concept for a promotional poster for my eco-friendly clothing brand. Can you help me brainstorm some unique ideas that will capture the essence of sustainability and style?"

Generating Color Palettes

Choosing the right color palette is crucial for creating a cohesive and visually appealing design. Chat-GPT can assist you in generating color palettes that complement your brand's identity and the message you want to convey.

Example Prompt: "My brand specializes in luxury skincare products. Can you suggest a color palette that conveys a sense of sophistication, elegance, and indulgence?"

Crafting Compelling Taglines and Headlines

Well-crafted taglines and headlines can make your visual content more engaging and effective. Chat-GPT can help you generate impactful taglines and headlines that capture your brand's essence and resonate with your target audience.

Example Prompt: "I'm creating a social media ad campaign for my online fitness coaching business. Can you help me come up with a powerful tagline that emphasizes the convenience and effectiveness of my coaching program?"

Providing Design Feedback

AI tools like Chat-GPT can offer constructive feedback on your visual content and graphic design, helping you identify areas for improvement and refine your creative vision.

Example Prompt: "I've designed a new logo for my coffee shop that features a minimalist coffee cup and the name of the shop in a modern font. Can you provide feedback on the design and suggest any improvements?"

Suggesting Font Pairings

Finding the perfect font pairing can elevate the overall aesthetic of your visual content. Chat-GPT can recommend font pairings that align with your brand's style and complement your design.

Example Prompt: "I'm designing a website for my digital marketing agency, and I need a professional yet modern font pairing. Can you suggest some options that will enhance the overall design?"

Creating Mood Boards

Mood boards can help you visualize and communicate the overall aesthetic of your brand or a specific project. Chat-GPT can help you curate images, colors, and design elements to create an inspiring mood board.

Example Prompt: "I'm working on a rebranding project for my travel agency, and I need to create a mood board that showcases a sense of adventure and luxury. Can you help me identify elements that would work well together in a mood board?"

Planning Visual Content Strategy

A well-planned visual content strategy is crucial for maintaining a consistent brand image across various platforms. Chat-GPT can help you develop a comprehensive content strategy that aligns with your brand's goals and appeals to your target audience.

Example Prompt: "I run a health food store and want to create a visual content strategy for our social media channels. Can you help me outline a plan that includes the types of visuals, posting frequency, and themes that will engage our audience?"

More AI Design Tools

1. Flair (https://flair.ai/): AI-powered online software for designing branded content, professional-looking graphics, and design elements easily.

2. Illustroke (https://illustroke.com/): AI-powered online software for creating vector images from text prompts.
3. Patterned (https://www.patterned.ai/): AI-powered online software for generating patterns for design.
4. Stockimg (https://stockimg.ai/): AI-powered online software for generating the perfect stock photo.
5. Looka (https://looka.com/): AI-powered online software for designing your brand identity.
6. Paintify AI (https://paintify.ai/): AI-powered online software that helps you create stunning artwork of various kinds easily

By embracing AI for visual content and graphic design, existing businesses and brands can elevate their creative assets, streamline their design process, and create a lasting impression on their target audience.

34

DOMINATE WITH SIMPLIFIED CODING

The Power of Chat-GPT Equals Coding Capabilities for Everyone

In today's technology-driven world, coding has become an essential skill for many individuals and businesses. Chat-GPT, an advanced AI language model, offers incredible support for people looking to harness the power of coding. This chapter will explore various coding tasks that Chat-GPT can perform to make your life easier, regardless of your coding experience.

1. Generating Code Snippets

Chat-GPT can generate code snippets in various programming languages such as Python, JavaScript, PHP, and more. By providing a description of the functionality you need, Chat-GPT can offer relevant code examples to implement in your projects.

1. Debugging and Code Review

Debugging can be a time-consuming process, but with Chat-GPT's assistance, you can identify errors or inefficiencies in your code. By providing the code snippet and explaining the issue, Chat-GPT can suggest possible solutions or improvements, helping you optimize your code.

1. API Integration

Integrating APIs into your projects can be challenging, especially if you are not familiar with the specific API's documentation. Chat-GPT can help you understand the API's functionality and generate code examples for integrating the API into your project, making the process more manageable.

1. Database Query Generation

Creating database queries can be complex, particularly for those unfamiliar with SQL or other query languages. Chat-GPT can help you generate queries based on your requirements, allowing you to retrieve, update, or manipulate data in your database with ease.

1. Web Development

Chat-GPT can assist with various aspects of web development, such as generating HTML, CSS, or JavaScript code snippets. By describing the design or functionality you want, Chat-GPT can provide relevant code examples to implement on your website.

1. Algorithm Creation

Developing algorithms can be a daunting task, but Chat-GPT can help you create algorithms for specific tasks or problems. By explaining the desired outcome, Chat-GPT can suggest algorithms or code snippets that can be adapted for your project.

1. Machine Learning Assistance

Machine learning is a rapidly growing field, and Chat-GPT can help you navigate the complexities of implementing machine learning algorithms in your projects. By providing a description of the problem you want to solve, Chat-GPT can suggest appropriate machine learning techniques and code examples to get you started.

Chat-GPT's coding capabilities can be a game-changer for both beginners and experienced developers alike. By utilizing this powerful AI tool, you can save time, improve your code's efficiency, and bring your projects to life with ease. Whether you need help with code generation, debugging, or algorithm creation, Chat-GPT is a valuable resource for anyone looking to excel in the world of coding.

35

CREATE A VIRTUAL SPOKESPERSON

Leverage AI to communicate effectively with your audience and enhance your brand's online presence

In today's digital age, having a virtual spokesperson can be a game-changer for businesses and brands. Chat-GPT, a powerful AI language model, can serve as an effective virtual spokesperson, enabling you to communicate efficiently with your target audience while saving time and resources. Let's explore specific ways to use Chat-GPT as a virtual spokesperson for your business or brand, along with example prompts to help you get started.

Answering Frequently Asked Questions (FAQs)

Chat-GPT can serve as a valuable resource for addressing common questions from your customers or audience, streamlining customer support, and enhancing user experience.

Example Prompt: "I run a software company specializing in project management tools. Can you help me draft answers to frequently asked questions about our software's features, pricing, and technical requirements?"

Drafting Press Releases and Announcements

Keep your audience informed of important updates, events, or product launches by using Chat-GPT to draft press releases and announcements that effectively communicate key information.

Example Prompt: "My company is launching a new line of eco-friendly home cleaning products. Can you help me draft a press release that highlights the benefits of our products and their positive impact on the environment?"

Generating Engaging Social Media Content

As a virtual spokesperson, Chat-GPT can help you create captivating social media content that resonates with your audience, strengthens your brand identity, and boosts engagement.

Example Prompt: "I own a boutique fitness studio, and I need engaging social media content for our Instagram account. Can you suggest some post ideas that showcase our classes, community, and overall wellness philosophy?"

Crafting Email Campaigns and Newsletters

Stay connected with your customers and prospects by using Chat-GPT to draft compelling email campaigns and newsletters that showcase your brand's offerings and share valuable insights.

Example Prompt: "I run an online clothing store, and I want to create a monthly newsletter that features new arrivals, fashion tips, and special promotions. Can you help me draft an engaging newsletter that highlights these elements?"

Responding to Customer Reviews and Feedback

Maintaining a positive online reputation is crucial for any business or brand. Chat-GPT can help you craft thoughtful responses to customer reviews and feedback, fostering a strong sense of community and demonstrating your commitment to customer satisfaction.

Example Prompt: "A customer left a review on our website expressing their disappointment with the delivery time of their order. Can you help me draft a professional and empathetic response addressing their concerns and outlining the steps we're taking to improve our delivery process?"

Creating Blog and Video Content

Position your brand as an industry authority by using Chat-GPT to develop informative blog articles and video scripts that educate your audience and showcase your expertise.

Example Prompt: "I'm a financial consultant, and I want to create a series of blog posts about personal finance management. Can you help me outline some topics and draft engaging articles that provide actionable tips and insights?"

Conducting Virtual Interviews and Q&A Sessions

Engage your audience in a dynamic way by using Chat-GPT to participate in virtual interviews or Q&A sessions. By providing informative and engaging answers, Chat-GPT can serve as an effective brand ambassador.

Example Prompt: "I'm hosting a live Q&A session for my health supplement brand. Can you help me draft responses to common questions about our products, their benefits, and recommended usage?"

Yes, Chat-GPT can serve as a virtual spokesperson, helping businesses and brands enhance their online presence, foster deeper connections with their audience, and streamline their communication efforts.

36

ELECTRIFY YOUR ONLINE STORE

Harness AI to improve user experience, increase conversions, and streamline customer support

In today's competitive ecommerce landscape, optimizing your website and online store is crucial for attracting and retaining customers. Chat-GPT, a powerful AI language model, can help you enhance various aspects of your ecommerce platform, ensuring a seamless user experience and driving sales. Let's explore specific ways to utilize Chat-GPT to improve your ecommerce website and online store, along with example prompts to help you get started.

Writing Compelling Product Descriptions

Well-written product descriptions can effectively showcase your offerings and convince potential customers to make a purchase. Chat-GPT can help you craft enticing product descriptions that highlight key features and benefits.

Example Prompt: "I sell artisanal candles made with natural ingredients. Can you help me write a compelling product description that emphasizes the unique scents, eco-friendly materials, and luxurious design?"

Generating Engaging Blog Content

High-quality blog content can position your brand as an industry authority, drive organic traffic, and increase customer engagement. Chat-GPT can help you create informative and captivating blog articles that resonate with your target audience.

Example Prompt: "I run an online store specializing in sustainable fashion. Can you help me draft a blog post that discusses the benefits of eco-friendly materials and highlights our commitment to sustainable practices?"

Crafting Persuasive Email Campaigns

Email marketing is a powerful tool for driving sales and nurturing customer relationships. Chat-GPT can help you develop compelling email campaigns that showcase your products, promotions, and brand story.

Example Prompt: "I'm launching a new collection of handcrafted jewelry in my online store. Can you help me create an engaging email campaign that promotes the collection and encourages customers to explore our new offerings?"

Answering Customer Inquiries

Chat-GPT can serve as an effective customer support resource, efficiently addressing common questions from your customers and streamlining the support process.

Example Prompt: "I own an online electronics store. Can you help me draft responses to frequently asked questions about our products' technical specifications, warranty information, and shipping policies?"

Optimizing Website Copy

Clear and engaging website copy is essential for providing a seamless user experience and effectively communicating your brand's value proposition. Chat-GPT can help you optimize your website copy, ensuring that it aligns with your brand's voice and resonates with your target audience.

Example Prompt: "I run an online store that sells gourmet coffee beans. Can you help me revise our homepage copy to better convey our passion for high-quality coffee and commitment to ethical sourcing practices?"

Curating Social Media Content

Social media is an invaluable tool for increasing brand visibility and engaging with your audience. Chat-GPT can help you generate captivating social media content that showcases your products and fosters a sense of community.

Example Prompt: "I have an online store that sells fitness equipment. Can you suggest some social media post ideas that highlight our products, share workout tips, and motivate our followers to lead a healthier lifestyle?"

Developing User Guides and Tutorials

User guides and tutorials can enhance your customers' experience with your products and provide valuable support. Chat-GPT can help you create comprehensive and easy-to-understand guides that ensure customer satisfaction.

Example Prompt: "I sell a line of smart home devices through my online store. Can you help me draft a user guide that walks customers through the setup process and provides troubleshooting tips for common issues?"

By setting sail with Chat-GPT, ecommerce businesses can improve various aspects of their online presence, delivering a seamless user experience and driving sales as they navigate the world of online entrepreneurship.

37

REVAMP YOUR BRICK-AND-MORTAR STORE

Harness AI to improve customer engagement, drive foot traffic, and streamline in-store operations

In an increasingly digital world, local brick-and-mortar stores must find innovative ways to compete and thrive. Chat-GPT, a powerful AI language model, can offer valuable support to help your local store increase revenue and enhance the customer experience. Let's explore specific ways to utilize Chat-GPT to boost your local store's performance, along with example prompts to help you get started.

Crafting Attention-Grabbing Advertising Copy

Well-crafted advertising copy can effectively promote your store and attract customers. Chat-GPT can help you create persuasive ad copy for print, radio, or online advertising that showcases your unique offerings and entices potential customers to visit your store.

Example Prompt: "I own a local bakery specializing in artisan bread and pastries. Can you help me write an engaging ad copy that highlights our freshly baked goods and invites customers to experience the cozy atmosphere of our store?"

Developing Localized Content Marketing

Content marketing can increase your store's visibility and establish your brand as a local authority. Chat-GPT can assist you in creating informative and engaging blog articles, newsletters, or social media posts that cater to your local audience and promote your store.

Example Prompt: "I run a neighborhood hardware store. Can you help me draft a blog post discussing common home maintenance issues in our area and how our store provides the necessary tools and expert advice to address them?"

Organizing In-Store Events and Promotions

In-store events and promotions can drive foot traffic and boost sales. Chat-GPT can help you brainstorm ideas for engaging events, workshops, or promotional offers that will attract customers and encourage them to spend more time (and money) at your store.

Example Prompt: "I own a local bookstore. Can you suggest some ideas for in-store events or promotions that will encourage people to visit and discover new books, engage with our staff, and participate in our community?"

Enhancing Customer Service and Support

Providing exceptional customer service is essential for retaining customers and generating positive word-of-mouth. Chat-GPT can help you draft employee training materials, customer support scripts, and responses to customer feedback, ensuring a high level of satisfaction.

Example Prompt: "I manage a local clothing boutique. Can you help me create a training guide for our employees that emphasizes the importance of personalized customer service and offers tips for assisting customers in finding the perfect outfit?"

Creating Compelling Store Signage and Displays

Eye-catching signage and displays can capture customers' attention and entice them to explore your store. Chat-GPT can help you generate ideas for effective store signage and display designs that showcase your products and create a welcoming atmosphere.

Example Prompt: "I run a local plant nursery. Can you suggest some ideas for store signage and display layouts that highlight our diverse plant selection, share care tips, and convey our passion for sustainable gardening practices?"

Designing Effective Loyalty Programs

Loyalty programs can encourage repeat business and foster long-term customer relationships. Chat-GPT can help you craft a loyalty program that rewards customers for their patronage, offers exclusive benefits, and strengthens their connection to your store.

Example Prompt: "I own a neighborhood coffee shop. Can you help me design a loyalty program that offers incentives for frequent visits, promotes our seasonal beverages, and encourages customers to engage with our community events?"

Implementing Targeted Email and SMS Marketing

Targeted email and SMS marketing can help you stay connected with your customers and keep them informed about upcoming sales, events, or new product arrivals. Chat-GPT can help you draft engaging and personalized marketing messages that resonate with your audience and drive them back to your store.

Example Prompt: "I own a local pet supply store. Can you help me create a targeted email campaign that informs our customers about our latest products, shares helpful pet care tips, and offers exclusive discounts for subscribers?"

Gathering and Analyzing Customer Feedback

Understanding your customers' needs and preferences is crucial for improving your store and increasing revenue. Chat-GPT can assist you in designing customer surveys or feedback forms that capture valuable insights and help you make data-driven decisions.

Example Prompt: "I manage a local grocery store. Can you help me draft a customer feedback survey that covers various aspects of our store, such as product selection, pricing, customer service, and store layout, to better understand our customers' preferences and areas for improvement?"

By using the power of Chat-GPT, local brick-and-mortar stores can implement innovative strategies to boost revenue, enhance customer engagement, and streamline in-store operations.

38

SUPERCHARGE YOUR SERVICE BUSINESS

Harness AI to improve customer engagement, generate leads, and streamline service operations

Local service businesses must continuously adapt and innovate to thrive in today's competitive market. Chat-GPT, a powerful AI language model, can offer valuable support to help your service business increase revenue and enhance customer satisfaction. Let's explore specific ways to utilize Chat-GPT to boost your local service business's performance, along with example prompts to help you get started.

Crafting Attention-Grabbing Advertising Copy

Well-crafted advertising copy can effectively promote your service business and attract potential customers. Chat-GPT can help you create persuasive ad copy for print, radio, or online advertising that highlights your unique offerings and entices customers to choose your services.

Example Prompt: "I own a local landscaping company. Can you help me write an engaging ad copy that showcases our expertise in creating beautiful outdoor spaces and highlights our commitment to sustainable practices?"

Developing Localized Content Marketing

Content marketing can increase your service business's visibility and establish your brand as a local authority. Chat-GPT can assist you in creating informative and engaging blog articles, newsletters, or social media posts that cater to your local audience and promote your services.

Example Prompt: "I run a local plumbing business. Can you help me draft a blog post discussing common plumbing issues in our area and how our experienced team provides fast, reliable, and affordable solutions?"

Generating Customer Testimonials and Case Studies

Positive customer testimonials and case studies can build trust and credibility for your service business. Chat-GPT can help you draft compelling testimonials and case studies that showcase your expertise and highlight the positive experiences of your satisfied customers.

Example Prompt: "I own a local home cleaning service. Can you help me create a case study that details a specific project where our team significantly improved a client's living space, resulting in a happier and more organized home environment?"

Enhancing Customer Service and Support

Providing exceptional customer service is essential for retaining customers and generating positive word-of-mouth. Chat-GPT can help you draft employee training materials, customer support scripts, and responses to customer feedback, ensuring a high level of satisfaction.

Example Prompt: "I manage a local car repair shop. Can you help me create a training guide for our employees that emphasizes the importance of clear communication, personalized service, and honest pricing?"

Creating Compelling Service Packages and Offers

Attractive service packages and special offers can entice customers to choose your business over competitors. Chat-GPT can help you develop tailored service packages and promotional offers that cater to your customers' needs and encourage them to use your services repeatedly.

Example Prompt: "I run a local hair salon. Can you suggest some ideas for service packages and promotional offers that appeal to various customer segments, such as first-time visitors, loyal clients, and special occasions?"

Designing Effective Referral Programs

Referral programs can encourage satisfied customers to spread the word about your service business, resulting in new leads and increased revenue. Chat-GPT can help you craft a referral program that rewards customers for their referrals and encourages them to share their positive experiences with friends and family.

Example Prompt: "I own a local photography studio. Can you help me design a referral program that offers incentives for existing customers to refer new clients, while also providing a discount for the referred customers?"

Implementing Targeted Email and SMS Marketing

Targeted email and SMS marketing can help you stay connected with your customers and keep them informed about upcoming promotions, service updates, or seasonal offers. Chat-GPT can help you draft engaging and personalized marketing messages that resonate with your audience and drive them to use your services.

Example Prompt: "I own a local fitness studio. Can you help me create a targeted email campaign that informs our customers about our latest classes, shares wellness tips, and offers exclusive discounts for subscribers?"

Gathering and Analyzing Customer Feedback

Understanding your customers' needs and preferences is crucial for improving your services and increasing revenue. Chat-GPT can assist you in designing customer surveys or feedback forms that capture valuable insights and help you make data-driven decisions.

Example Prompt: "I manage a local catering business. Can you help me draft a customer feedback survey that covers various aspects of our service, such as food quality, presentation, punctuality, and customer service, to better understand our customers' preferences and areas for improvement?"

Streamlining Appointment Scheduling and Reminders

Efficient appointment scheduling and timely reminders can improve customer satisfaction and reduce no-shows. Chat-GPT can help you draft clear and friendly appointment confirmation emails or SMS messages, as well as reminders that keep your customers informed and ensure smooth service delivery.

Example Prompt: "I run a local dental practice. Can you help me create an appointment confirmation email and a reminder message that provide all the necessary information, while also conveying our commitment to patient care and comfort?"

With Chat-GPT at their disposal, local service businesses can implement innovative strategies to boost revenue, enhance customer engagement, and streamline service operations.

39

STREAMLINE YOUR BUSINESS

Use AI to increase efficiency, enhance communication, and drive growth

In today's fast-paced business world, maximizing efficiency and reducing operational costs are crucial for maintaining a competitive edge. Chat-GPT, a powerful AI language model, offers a wide range of applications that can help businesses save time, streamline processes, and ultimately increase profits. Let's explore specific ways to utilize Chat-GPT to optimize your business operations, along with example prompts to help you get started.

Automating Customer Support

Providing timely and efficient customer support can significantly impact customer satisfaction and loyalty. Chat-GPT can be integrated into chatbots or email systems to answer frequently asked questions, resolve common issues, or guide customers through troubleshooting processes.

Example Prompt: "Create a chatbot response script that

addresses common questions about our product's installation process, warranty information, and troubleshooting tips."

Streamlining Internal Communication

Clear and effective communication among team members is essential for smooth operations and collaboration. Chat-GPT can help draft emails, create meeting agendas, and generate concise summaries of lengthy documents, saving time and ensuring efficient communication across your organization.

Example Prompt: "Please create a concise summary of our latest quarterly financial report, highlighting key performance indicators and areas of improvement."

Enhancing Content Creation and Marketing

High-quality content is crucial for capturing your audience's attention and promoting your brand. Chat-GPT can help generate engaging blog posts, persuasive ad copy, and attention-grabbing social media captions, reducing the time and effort required to produce impactful marketing materials.

Example Prompt: "Craft a persuasive and engaging email campaign promoting our upcoming product launch, highlighting its unique features and benefits."

Conducting Market Research and Analysis

Understanding market trends, customer preferences, and competitor strategies is critical for making informed business decisions. Chat-GPT can help gather and analyze relevant data, generate market research reports, and identify opportunities for growth.

Example Prompt: "Analyze the current market trends in the sustainable fashion industry and identify three emerging opportunities for our brand to capitalize on."

Streamlining Human Resources Processes

Efficient HR processes are essential for attracting, retaining, and managing talent within your organization. Chat-GPT can help draft job descriptions, screen resumes, and automate repetitive HR tasks, allowing your HR team to focus on more strategic initiatives.

Example Prompt: "Create a job description for a marketing manager position, outlining key responsibilities, required qualifications, and desired skills."

Facilitating Product Development

Incorporating customer feedback and staying abreast of industry trends are vital for developing products that meet consumer needs. Chat-GPT can help analyze customer reviews, generate product improvement suggestions, and brainstorm innovative product ideas.

Example Prompt: "Based on recent customer feedback, suggest three potential improvements to our mobile app's user interface and functionality."

Optimizing Business Processes and Workflow

Efficient workflows and processes are critical for maximizing productivity and reducing operational costs. Chat-GPT can help identify bottlenecks, suggest process improvements, and create detailed SOPs (standard operating procedures) to ensure consistent performance across your organization.

Example Prompt: "Review our current customer service process and suggest three potential improvements to reduce response time and increase customer satisfaction."

PART IV

AMPLIFY YOUR CREATIVITY

40

BLOG LIKE A BOSS

Starting a blog can be an exciting way to share your passion and knowledge with the world. But did you know that you can also turn your blog into a profitable venture? Let's explore various methods that bloggers can use to generate income from their blogs.

Affiliate Marketing

One of the most popular ways to monetize a blog is through affiliate marketing. Bloggers can partner with companies or affiliate networks to promote products or services on their blogs. In return, they earn a commission for every sale made through their unique referral links. To succeed in affiliate marketing, choose products that align with your blog's niche and are genuinely helpful to your audience.

Sponsored Posts and Advertisements

Another way to earn income from your blog is by collaborating with brands to create sponsored content or display advertisements.

Sponsored posts involve creating content that promotes a specific product or service, while display advertisements are banner ads placed on your blog. To attract sponsors and advertisers, focus on building your blog's credibility and growing your audience.

Selling Digital Products

Bloggers can also create and sell digital products, such as eBooks, online courses, webinars, or printables. This method allows you to leverage your expertise and generate passive income. Ensure that the digital products you create provide value to your audience and help solve a problem or fulfill a need.

Offering Consulting or Coaching Services

If you're an expert in your field, you can offer consulting or coaching services to your audience. Many bloggers successfully monetize their blogs by providing personalized advice, mentoring, or training to their readers. To get started, create a dedicated page on your blog outlining your services and showcasing your expertise.

Subscription-based Content or Memberships

Another way to generate income from your blog is by offering premium, subscription-based content or memberships. This can include exclusive articles, podcasts, or videos available only to paying subscribers. To make this option appealing, ensure that your premium content offers significant value and benefits to your subscribers.

Donations and Crowdfunding

While not as common, some bloggers opt to generate income through donations or crowdfunding. This method involves asking your audience to support your work financially, either through one-time donations or recurring payments. Platforms like Patreon, Ko-fi, or Buy Me a Coffee can help facilitate this process.

Starting a blog can be a rewarding and potentially lucrative endeavor. By exploring various income-generating methods and focusing on providing value to your audience, you can turn your passion project into a thriving online business. With dedication, creativity, and persistence—and using some of the many AI resources available now—the possibilities for earning income from your blog are endless.

41

AMPLIFY YOUR BLOG'S REVENUE

Harness AI to create engaging content, optimize SEO, and streamline your blogging workflow

As a blogger, generating engaging content, increasing your blog's reach, and monetizing your platform can be time-consuming and challenging. Chat-GPT, a powerful AI language model, can offer valuable support to help you create captivating content, optimize SEO, and ultimately increase your blog's revenue. Let's explore specific ways to utilize Chat-GPT to elevate your blogging game, along with example prompts to help you get started.

Generating High-Quality Content Ideas

Coming up with fresh, unique, and relevant content ideas can be difficult. Chat-GPT can help you brainstorm and create a list of compelling blog post ideas that cater to your target audience and keep them engaged.

Example Prompt: "My blog is about eco-friendly living. Can you suggest 10 engaging blog post ideas that cover various aspects of sustainable living and provide actionable tips for my readers?"

Crafting Engaging Blog Post Titles

An attention-grabbing blog post title can significantly impact your click-through rate and overall blog traffic. Chat-GPT can help you create captivating and share-worthy blog post titles that pique readers' curiosity and encourage them to explore your content.

Example Prompt: "I've written a blog post about the health benefits of a plant-based diet. Can you suggest five eye-catching titles that would encourage readers to click and read the post?"

Writing and Editing Blog Posts

Creating well-written, informative, and engaging blog posts can be time-consuming. Chat-GPT can assist you in drafting and editing your blog posts, ensuring they are coherent, well-structured, and error-free.

Example Prompt: "I want to write a blog post about the top 10 travel destinations for digital nomads. Can you help me create an engaging and informative article that highlights the main attractions, cost of living, and digital infrastructure for each destination?"

Enhancing SEO and Keyword Integration

Optimizing your blog for search engines is crucial for increasing your online visibility and organic traffic. Chat-GPT can help you identify relevant keywords and naturally integrate them into your blog posts without compromising readability.

Example Prompt: "My blog post is about zero-waste kitchen tips. Can you suggest 5-7 relevant keywords for this topic, and help me incorporate them into the text without making it sound forced or unnatural?"

Creating Engaging Social Media Content

Promoting your blog on social media platforms can attract new readers and increase your blog's reach. Chat-GPT can help you craft engaging social media posts, captions, and hashtags that resonate with your target audience and encourage them to visit your blog.

Example Prompt: "I've just published a blog post on the benefits of meditation for stress relief. Can you help me create three different social media posts to promote this article on Facebook, Twitter, and Instagram?"

Designing Effective Email Campaigns

Email marketing can help you build a loyal readership, promote your latest content, and generate revenue through affiliate marketing or sponsored content. Chat-GPT can assist you in crafting compelling email newsletters and campaigns that keep your subscribers engaged and drive traffic back to your blog.

Example Prompt: "I want to send a monthly newsletter to my subscribers that includes a summary of my latest blog posts, personal updates, and exclusive content. Can you help me draft an engaging newsletter that encourages readers to check out my recent posts?"

Streamlining Your Blogging Workflow

Chat-GPT can also help you manage your blogging tasks more efficiently by drafting outlines, creating content calendars, and setting goals for your blog.

Example Prompt: "I want to plan my blog content for the next three months. Can you help me create a content calendar that includes a mix of evergreen articles, seasonal topics, and guest posts, with specific publishing dates and deadlines?"

Monetizing Your Blog

Generating revenue from your blog can be achieved through various methods, such as sponsored posts, affiliate marketing, or selling digital products. Chat-GPT can help you draft persuasive pitches to potential sponsors, craft affiliate product reviews, or even develop outlines for ebooks or online courses.

Example Prompt: "I'd like to pitch a sponsored post idea to a sustainable fashion brand. Can you help me write a professional and persuasive email that highlights the benefits of partnering with my eco-friendly living blog and presents a unique content idea that aligns with their brand values?"

Engaging with Your Audience

Building a loyal readership requires consistent engagement and interaction with your audience. Chat-GPT can assist you in drafting thoughtful responses to blog comments or help you create engaging content for Q&A sessions, webinars, or live streams.

Example Prompt: "I'm hosting a live Q&A session about sustainable living on my blog. Can you help me prepare a list of 10 engaging questions that cover various aspects of eco-friendly living, and also provide concise and informative answers to spark discussion during the session?"

By learning and using the power of Chat-GPT, bloggers can create engaging content, optimize SEO, streamline their workflow, and ultimately increase their blog's reach and revenue. Embrace the potential of AI technology to help your blog thrive in today's competitive online landscape.

42

TOP FIVE BLOGGING PLATFORMS

WordPress (Self-Hosted)

Pros:

- Complete control and customization: With self-hosted WordPress, you have the flexibility to customize your website's design, features, and functionality.
- Extensive plugin ecosystem: Choose from thousands of free and premium plugins to enhance your blog's performance and user experience.
- SEO-friendly: WordPress is built with search engine optimization in mind, making it easier for your content to rank higher in search results.
- Large community and support: WordPress has a massive community and extensive documentation, ensuring that help is readily available.

Cons:

- Requires some technical knowledge: Setting up a self-hosted WordPress blog may be intimidating for beginners, as it involves purchasing a domain name and web hosting.
- Maintenance and updates: You are responsible for maintaining and updating your website, which may require some technical know-how.

Blogger

Pros:

- Easy to use: Blogger offers a user-friendly interface, making it simple for beginners to create and publish blog posts.
- Free and reliable: Blogger is a free platform backed by Google, ensuring stability and reliability.
- Google integration: Blogger seamlessly integrates with other Google services, such as Google Analytics and AdSense.

Cons:

- Limited customization: Blogger's design and feature options are more limited compared to other platforms like WordPress.
- Less control: Google owns Blogger, meaning they can shut down your blog at any time if they find it violates their terms of service.

Wix

Pros:

- Drag-and-drop website builder: Wix's user-friendly interface allows you to design your blog with ease, even if you have no coding knowledge.
- Numerous templates: Choose from a wide selection of professionally designed templates to create an attractive blog.
- Free plan available: Wix offers a free plan, making it a budget-friendly option for new bloggers.

Cons:

- Limited SEO capabilities: Wix has improved its SEO features, but it still lags behind other platforms like WordPress.
- Limited scalability: Wix may not be suitable for large, complex websites or those with advanced features.

Squarespace

Pros:

- Beautiful design templates: Squarespace is known for its visually stunning templates, making it an ideal choice for bloggers focused on aesthetics.
- Ease of use: Squarespace's intuitive interface makes it easy to create and manage your blog.
- Integrated e-commerce: Squarespace offers built-in e-commerce capabilities, allowing you to sell products directly through your blog.

Cons:

- Limited flexibility: Squarespace's customization options are more limited compared to WordPress, making it less suitable for advanced users.

- Higher cost: Squarespace's pricing plans are more expensive than other platforms, particularly for those requiring e-commerce features.

Medium

Pros:

- Simple and clean interface: Medium's focus is on content, making it an excellent platform for writers who prioritize text over design.
- Built-in audience: Medium has a large, engaged community, which can help expose your content to new readers.
- No technical skills required: Medium is easy to use, with no need to worry about hosting, domain names, or website maintenance.

Cons:

- Limited control: Medium offers minimal customization options and control over your blog's design and features.
- Monetization restrictions: Medium does not support display ads, limiting your ability to monetize your blog through advertising.

43

WORDPRESS AND AI

I'm a longtime Wordpress user, and given its prominence and versatility, that's the software I usually recommend. As such, I'll use Wordpress in the examples below while explaining some of the way to boost your blogging revenue with AI and Chat-GPT. If you prefer a different blogging software, simply include its name in a prompt and ask Chat-GPT to explore similar use cases.

Having a WordPress blog connected to an online business or brand provides numerous benefits, and Chat-GPT can play a significant role in enhancing the blog's effectiveness. Some of the benefits of a WordPress blog include:

1. Improved SEO: A regularly updated blog with relevant, high-quality content can help improve search engine rankings, driving more organic traffic to your business website.
2. Enhanced Customer Engagement: A blog allows you to share valuable information, tips, and insights with your audience, fostering a deeper connection with your customers and building trust.

3. Increased Brand Awareness: Blogging helps showcase your brand's personality, expertise, and values, establishing your business as an industry thought leader and attracting new customers.
4. Content Marketing: A blog is an essential component of any content marketing strategy, providing a platform to share and promote content across various channels, such as social media and email marketing.
5. Lead Generation: By offering valuable content and implementing effective calls-to-action, a blog can serve as a lead generation tool, capturing potential customers' contact information for future marketing efforts.

How Chat-GPT can help with your WordPress blog:

1. Content Creation: Chat-GPT can help you generate engaging, relevant, and well-written blog post ideas, titles, outlines, and even complete articles based on your chosen topics or keywords.
2. SEO Optimization: Chat-GPT can assist with keyword research, suggesting keywords to target in your content to improve search engine rankings and drive more organic traffic.
3. Editing and Proofreading: Chat-GPT can help review and edit your blog posts for grammar, syntax, and overall readability, ensuring a polished and professional final product.
4. Social Media Promotion: Chat-GPT can generate compelling social media captions, hashtags, and updates to promote your blog content across various platforms, driving traffic and engagement.
5. Blog Management: Chat-GPT can help you brainstorm and plan your editorial calendar, ensuring consistent

publishing and a well-rounded content mix that caters to your audience's interests.

All in all, AI and especially Chat-GPT can help you create a more effective WordPress blog, especially one that can add to the growth of an online business or brand.

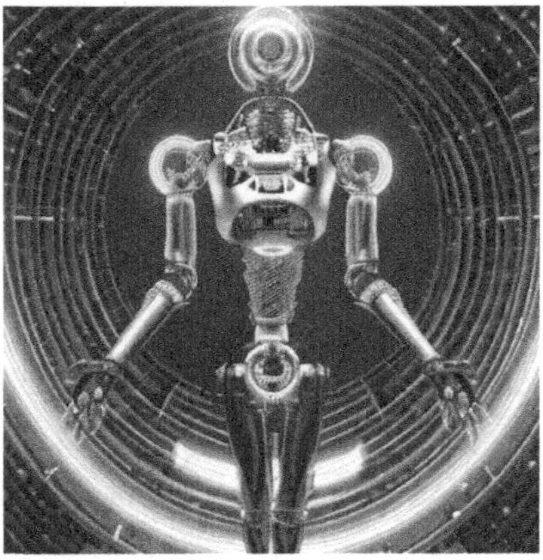

Start A Blog – Step by Step

Starting a WordPress blog is a straightforward process. Here are the steps to get your blog up and running:

1. Choose a domain name: A domain name is your website's address on the internet, such as example.com. Try to pick a short, memorable, and relevant domain name that reflects your blog's niche or topic.
2. Register your domain: Once you've chosen a domain name, you'll need to register it through a domain registrar, like GoDaddy, Namecheap, or Google Domains. This will typically require an annual fee.
3. Select a hosting provider: Web hosting is a service that stores your website files and makes them accessible on the internet. Choose a reliable hosting provider, such as

SiteGround, Bluehost, or WP Engine, that offers WordPress-specific hosting plans.

4. Install WordPress: Most hosting providers offer a one-click WordPress installation through their control panel. Log in to your hosting account, navigate to the control panel, and follow the instructions to install WordPress.
5. Configure WordPress settings: After installation, log in to your WordPress admin dashboard (usually found at yourdomain.com/wp-admin). Here, you can configure your site's settings, such as site title, tagline, and permalink structure.
6. Choose a theme: WordPress themes dictate your blog's appearance and layout. Select a theme that suits your niche and personal style. You can find free themes in the WordPress theme repository or purchase premium themes from marketplaces like ThemeForest or independent theme developers.
7. Customize your theme: After installing a theme, customize it to match your branding and preferences. Most themes offer customization options, like color schemes, fonts, and header styles, which can be accessed through the WordPress customizer (Appearance > Customize).
8. Install plugins: Plugins extend the functionality of your WordPress site. Some essential plugins to consider are Yoast SEO (for search engine optimization), Akismet (for spam protection), and Jetpack (for site performance and security).
9. Create content: Start by creating essential pages, such as an "About" page and a "Contact" page. Then, begin drafting and publishing blog posts. Make sure to create high-quality, engaging, and valuable content that appeals to your target audience.
10. Promote your blog: Share your blog posts on social media platforms, engage in relevant online

communities, and use email marketing to drive traffic to your blog. As your blog grows, consider employing SEO and content marketing strategies to further boost your online presence.

By following these steps, you'll have a WordPress blog up and running, ready to attract readers and contribute to your online business or personal brand's growth.

44

PRODUCE PLUGINS WITH AI

Artificial Intelligence (AI) has revolutionized the way businesses operate, and Chat-GPT is no exception. With its advanced natural language processing capabilities, Chat-GPT can significantly assist in generating ideas for WordPress plugins tailored to your niche scented candle business. Let's explore the benefits of using Chat-GPT for plugin creation and provide examples of plugins that can enhance your blog.

Benefits of Using Chat-GPT for WordPress Plugin Creation

1. Time-saving: Chat-GPT's AI capabilities can quickly generate numerous ideas for potential plugins, saving you valuable time that would be spent brainstorming.
2. Creative solutions: Chat-GPT can provide fresh perspectives and innovative ideas that may not have been considered otherwise.
3. Improved user experience: By creating plugins specifically tailored to your brand or business, you can provide a unique and engaging experience for your blog visitors.

4. Enhanced functionality: Chat-GPT can help identify potential gaps in your blog's current functionality, providing plugin ideas to address those needs.
5. Personalized recommendations: Chat-GPT can be prompted to provide plugin ideas that align with your brand and target audience, ensuring a cohesive and consistent user experience.

I prompted Chat-GPT to "list some ideas of Wordpress plug-ins that you can create for my blog connected to my scented candle business." Chat-GPT quickly responded with the following:

Plugin Ideas for Your Scented Candle Blog

1. Candle Scent Finder: This plugin would allow visitors to answer a series of questions about their scent preferences, generating personalized scent recommendations based on their input.
2. Seasonal Scent Calendar: This plugin could showcase the most popular scents for each season, helping visitors discover new fragrances and driving sales of seasonal candles.
3. Candle Care Tips: A plugin that displays helpful tips and tricks for proper candle care, including burning, wick trimming, and storage, to help customers maximize the longevity of their candles.
4. Interactive Candle Gallery: This plugin would enable visitors to view high-resolution images and descriptions of various candle types, scents, and sizes, providing an immersive browsing experience.
5. Scent Pairing Guide: A plugin that suggests complementary scents for customers looking to create a multisensory experience in their homes.

6. Candle Customizer: Allow customers to design their own candles by choosing the size, color, and fragrance, adding a personal touch to their purchase.
7. Candle Subscription Box Builder: This plugin could enable visitors to create a custom subscription box by selecting their preferred scents, with the option for recurring delivery.
8. Gift Guide: A plugin that generates personalized gift recommendations based on the recipient's interests and preferences, making gift-giving easy for your customers.
9. Scent Memory Stories: A community-driven plugin that allows users to share their scent memories and experiences, fostering engagement and building a sense of community among your audience.
10. Candle-Making Tutorials: A plugin that features step-by-step guides and video tutorials for visitors interested in making their own scented candles at home.

Aside from the ideas it can generate, Chat-GPT is well-suited to help you with the development of innovative and engaging WordPress plugins for your blog. By providing a unique user experience and addressing your audience's specific needs through AI-enhanced blogging, you can amplify your voice, attract more visitors, and drive sales for a business.

Wordpress Plugins Step by Step

Following up with the first idea Chat-GPT generated, I prompted it with the following:

"Create a Candle Scent Finder plugin for Wordpress as you described in your first example and provide me with the necessary codes and instructions."

Chat-GPT responded that it would guide me through the general steps to create a Candle Scent Finder plugin for WordPress, adding that I would need "basic knowledge of PHP, HTML, CSS, and JavaScript, as well as experience with the WordPress plugin structure."

The instructions generated by Chat-GPT are as follows:

Here's a high-level overview of the steps you'll need to take:

1. Set up your plugin folder and files:
2. Create a new folder in your WordPress installation's **/wp-content/plugins/** directory, e.g., **candle-scent-finder.**
3. Inside the folder, create a main PHP file, e.g., **candle-scent-finder.php.**
4. Add a plugin header in the PHP file to provide information about the plugin, such as its name, author, version, and description.
5. Create a shortcode to display the Candle Scent Finder form:
6. In **candle-scent-finder.php**, create a function that generates the HTML for the form. This form should include questions about the user's scent preferences and a submit button.
7. Register the shortcode using **add_shortcode()** in your main PHP file. This will enable you to insert the form on any page or post by using the shortcode **[candle_scent_finder]**.

8. Create JavaScript logic for handling the form submission and generating scent recommendations:
9. Enqueue a JavaScript file in your plugin using **wp_enqueue_script().**
10. In the JavaScript file, add an event listener for the form submission.
11. When the form is submitted, prevent the default behavior and process the user's input to generate personalized scent recommendations.
12. Display the recommendations on the same page as the form, either by appending them to the DOM or updating an existing element.
13. Style the form and recommendations with CSS:
14. Enqueue a CSS file in your plugin using **wp_enqueue_style().**
15. In the CSS file, add custom styles for the form elements, recommendations, and any other components you want to style.
16. Test the plugin:
17. Activate the plugin in your WordPress dashboard.
18. Add the shortcode **[candle_scent_finder]** to a page or post.
19. Test the form to ensure it's working correctly and generating accurate scent recommendations.

While I can't provide the actual code, I hope this high-level overview helps guide you through the process of creating a Candle Scent Finder plugin for your WordPress site.

Although Chat-GPT said that couldn't provide the actual code, it apparently isn't always as self-aware as it may usually seem. I followed up by asking it to give me the code for the plugin. This is the prompt I used:

"Create the code for the plugin 'candle-scent-finder.php' with a function that generates HTML for the form and include common examples of scent preferences, and also include code for a submit button for the function."

Chat-GPT provided me with the code as seen in the following screenshot, and, furthermore, it continued to help me fine-tune the code while I tested it in my Wordpress blog until the Plugin was to my liking.

45

UNLOCK PASSIVE INCOME

Use Chat-GPT to create your side-hustle

In today's fast-paced world, finding ways to generate passive income has become increasingly popular. Chat-GPT, an advanced AI language model, offers various opportunities to create multiple streams of passive income. Let's explore some of the best ways to utilize Chat-GPT to generate passive income and pave the way to financial freedom.

Create and sell digital products

One of the most effective ways to generate passive income with Chat-GPT is by creating digital products. Use the AI to help you develop ebooks, online courses, or other educational materials on your chosen topic. For instance, you could ask Chat-GPT to help you outline an ebook on personal finance, and then use the generated content as the basis for your product. Once your digital product is complete, sell it on platforms like Amazon Kindle, Gumroad, or Teachable.

Start a niche blog or website

Chat-GPT can be an invaluable tool for creating content for a niche blog or website. By using the AI to generate articles and blog posts, you can quickly build up a content library on your chosen subject. Over time, your blog can attract organic traffic from search engines, allowing you to monetize your site through advertising, affiliate marketing, or sponsored content.

Develop a chatbot or AI-powered service

Leverage the capabilities of Chat-GPT to develop a chatbot or AI-powered service for businesses or individuals. You can create a subscription-based service offering tailored solutions, such as AI-generated social media content or customer support chatbots. Once your service is up and running, you'll earn passive income through monthly subscription fees from your clients.

Freelance writing and ghostwriting

While not entirely passive, Chat-GPT can significantly streamline your writing process and boost your productivity as a freelance writer or ghostwriter. You can use the AI to generate content quickly, allowing you to take on more clients and projects. As you establish a reputation for quality and efficiency, you can increase your rates and outsource more of the writing process to the AI, transforming this income stream into a more passive

Write and publish books

With the help of Chat-GPT, writing and publishing your books, whether fiction or non-fiction, becomes a more manageable task. The AI can help you with plot development, character creation, or even writing entire chapters. Once your book is complete, you can self-publish on platforms like Amazon Kindle Direct Publishing and earn royalties on every sale.

Create AI-generated art or design

If you have a flair for art or design, Chat-GPT can be used in combination with other AI-powered tools to create unique visual content. You can sell your AI-generated artwork as digital downloads, prints, or merchandise on platforms like Etsy, Redbubble, or Society6, generating passive income from your creations.

By harnessing the power of Chat-GPT, you can explore various avenues to generate passive income. From creating digital products to offering AI-powered services, the opportunities are vast and diverse. Start experimenting with Chat-GPT today and unlock your passive income potential.

46

SENSATIONAL SIDE-HUSTLES

Start a Profitable Online Course with Chat-GPT

The online education industry has experienced tremendous growth in recent years. Creating and selling online courses has become a popular and profitable business for entrepreneurs and subject matter experts. In this chapter, we'll explore how Chat-GPT can help you create engaging, high-quality online courses and provide a list of the top 5 most lucrative online course ideas.

1. Course Content Creation

Chat-GPT can assist in generating high-quality course content, including written materials, scripts for video lectures, and engaging quizzes. By providing a course outline or a list of topics you want to cover, Chat-GPT can help you craft comprehensive and informative course content.

1. Course Outline and Structure

Designing a well-structured course outline is essential for delivering a great learning experience. Chat-GPT can help you create a logical course outline by suggesting the ideal sequence of topics, lessons, and assessments based on your target audience and learning objectives.

1. Marketing and Sales Copy

To sell your online courses effectively, you need compelling marketing copy. Chat-GPT can help you write persuasive sales copy, landing page content, and email sequences to promote your courses and attract potential students.

1. Content Optimization and SEO

For your online courses to gain visibility and traction, you need to optimize your course content for search engines. Chat-GPT can assist in generating SEO-friendly content, such as keyword-rich course descriptions and blog articles, to improve your search engine rankings and drive organic traffic to your course offerings.

1. Customer Support and Q&A

Maintaining a high level of customer support is crucial for the success of your online courses. Chat-GPT can help you create a comprehensive FAQ section and provide prompt responses to student queries, enhancing the overall learning experience and increasing student satisfaction.

Top 5 Most Lucrative Online Course Ideas

1. Digital Marketing: With businesses constantly seeking ways to improve their online presence, a course on

digital marketing strategies, including social media, SEO, and content marketing, is in high demand.

2. Web and App Development: As technology evolves, there's a growing need for skilled web and app developers. Offering a course on programming languages, frameworks, and development tools can attract aspiring developers and professionals looking to upskill.
3. Personal Finance and Investing: With increasing interest in personal finance management and investment strategies, a course covering budgeting, saving, investing, and retirement planning can be highly profitable.
4. Health and Wellness: As people become more conscious of their health, courses on nutrition, fitness, stress management, and mental health can generate significant revenue.
5. Graphic Design and Multimedia: With the growing importance of visual content in marketing, a course on graphic design, video editing, and multimedia production can be a lucrative venture.

Now, more than ever before, Chat-GPT can help you create engaging, high-quality online courses and tap into the thriving e-learning market. From course content creation to marketing and customer support, Chat-GPT offers a range of capabilities to help you develop and sell profitable online courses, ensuring success in this lucrative industry.

Create Digital Products with Chat-GPT

The digital products market has seen exponential growth in recent years, as more and more people seek to monetize their skills and knowledge. In this chapter, we'll explore how Chat-GPT can help you create engaging, high-quality digital products and provide a list of the top 5 most lucrative digital product ideas.

1. Digital Product Idea Generation

Coming up with innovative and in-demand digital product ideas can be challenging. Chat-GPT can help you brainstorm and refine your ideas based on your expertise, target audience, and market demand.

1. Content Creation and Editing

Chat-GPT can assist in generating high-quality content for your digital products, including ebooks, reports, guides, and templates. By providing an outline or a list of topics you want to cover, Chat-GPT can help you craft comprehensive and informative content that meets the needs of your target audience.

1. Design and Visual Elements

Visual appeal is crucial for the success of your digital products. Chat-GPT can help you create design concepts, color schemes, and typography choices, enhancing the overall look and feel of your products.

1. Marketing and Sales Copy

To sell your digital products effectively, you need compelling marketing copy. Chat-GPT can help you write persuasive sales copy, landing page content, and email sequences to promote your products and attract potential customers.

1. Customer Support and Q&A

Maintaining a high level of customer support is essential for the success of your digital products. Chat-GPT can help you create a comprehensive FAQ section and provide prompt responses to customer queries, enhancing their overall experience and increasing customer satisfaction.

Top 5 Most Lucrative Digital Product Ideas

1. Ebooks: Writing ebooks or guides on in-demand topics, such as self-improvement, personal finance, or business strategies, can generate significant revenue.
2. E-Guides: Earn profits from selling electronically-distributed guides (for example, in a password-protected PDF) that demystify complex topics or provide readers with instructions on popular skillets, such as DIY projects, software guides, business snapshots.
3. Premium Content Subscriptions: Offering a subscription service that provides exclusive content, such as articles, videos, or podcasts, can be a profitable venture, especially if you have a loyal audience or a niche market.
4. Software and Mobile Apps: Developing useful software tools, such as productivity apps, project management tools, or graphic design applications, can cater to the needs of various industries and generate substantial income.
5. Digital Art and Design Assets: Creating digital assets, such as stock photos, vector illustrations, and design templates, can be a lucrative business, as many individuals and businesses require these resources for their projects.

Chat-GPT is the perfect tool to help you create engaging, high-quality digital products that cater to the needs of your target audience.

From idea generation to content creation, design, and marketing, Chat-GPT offers a range of capabilities to help you become an expert in a wide range of fields and then develop and sell profitable digital products that provide valuable information to your audience.

47

SUPERB SOCIAL INFLUENCE

Harness the power of AI to create engaging content, optimize your posting strategy, and monetize your influence

As a social media influencer, creating captivating content, growing your audience, and monetizing your influence can be challenging. Chat-GPT, a powerful AI language model, can provide valuable support in generating engaging content, optimizing your posting strategy, and ultimately increasing your reach, followers, and income. Let's explore specific ways to utilize Chat-GPT to take your influencer game to the next level, along with example prompts to help you get started.

Generating Fresh and Engaging Content Ideas

Staying relevant and keeping your followers engaged requires a constant stream of unique and captivating content. Chat-GPT can help you brainstorm creative content ideas that cater to your target audience's interests and preferences.

Example Prompt: "I am a fitness influencer looking to create more engaging content for my Instagram feed. Can you suggest 10 content ideas that include a mix of workout tutorials, nutrition tips, and motivational posts?"

Crafting Captivating Captions and Hashtags

An eye-catching caption and strategic use of hashtags can significantly impact your engagement rate and overall reach. Chat-GPT can help you create compelling captions and identify relevant hashtags that encourage your followers to interact with your content and attract new audience members.

Example Prompt: "I'm posting a photo of my latest vegan meal on Instagram. Can you help me write a captivating caption that describes the dish and shares the health benefits, along with five relevant hashtags to increase its visibility?"

Developing Long-Form Content for Blogs or YouTube

Long-form content, such as blog posts or YouTube videos, can help you establish authority and credibility within your niche. Chat-GPT can assist you in outlining and scripting engaging long-form content that resonates with your audience and provides value.

Example Prompt: "I'm a beauty influencer planning to create a YouTube video on skincare routines for different skin types. Can you help me draft a video script that covers essential skincare steps, product recommendations, and tips for each skin type?"

Enhancing Your Posting Strategy and Timing

Optimizing your posting strategy and timing can significantly improve your content's visibility and engagement. Chat-GPT can help you analyze your audience's preferences and develop a tailored posting schedule that maximizes your reach and impact.

Example Prompt: "I'm a travel influencer looking to optimize my posting schedule on Instagram, Twitter, and Facebook. Can you help me create a posting strategy that accounts for peak engagement times, content variety, and platform-specific best practices?"

Designing Collaborations and Partnerships

Collaborating with fellow influencers or partnering with brands can boost your reach, credibility, and income. Chat-GPT can help you draft partnership proposals, design collaborative content ideas, and craft persuasive pitches to potential sponsors.

Example Prompt: "I'm a fashion influencer interested in collaborating with a sustainable clothing brand. Can you help me write a proposal email that highlights the benefits of partnering with me and presents a unique content idea that aligns with the brand's values and goals?"

Monetizing Your Influence

Generating income as a social media influencer can be achieved through various methods, such as sponsored posts, affiliate marketing, or selling digital products. Chat-GPT can help you create persuasive sponsored content, draft authentic product reviews, or develop outlines for ebooks or online courses.

Example Prompt: "I'm a wellness influencer looking to promote a meditation app through a sponsored Instagram post. Can you help me write a genuine and engaging caption that highlights the app's features and benefits, while also providing a call-to-action for my followers to download the app?"

Engaging with Your Audience

Building a loyal and engaged following requires consistent interaction with your audience. Chat-GPT can assist you in crafting thoughtful responses to comments or direct messages, and help you create engaging content for Q&A sessions, webinars, or live streams.

Example Prompt: "I'm hosting a live Q&A session on Instagram about my fitness journey. Can you help me prepare a list of 10 engaging questions that cover various aspects of my fitness journey, and also provide concise and informative answers to spark discussion during the session?"

Analyzing Your Performance and Setting Goals

Monitoring your performance and setting achievable goals are essential for growing your influence and income. Chat-GPT can help you analyze your engagement metrics and develop a strategic plan to improve your reach, audience growth, and monetization efforts.

Example Prompt: "I'd like to create a 3-month plan to grow my Instagram following and increase my income from sponsored posts. Can you help me set specific, measurable goals and outline actionable steps I can take to achieve these goals?"

With Chat-GPT, social media influencers can create engaging content, optimize their posting strategy, and monetize their influence more effectively. AI technology will help elevate your social media presence and maximize your income in today's competitive influencer landscape.

Social Media Management: Step by Step

Using Chat-GPT for social media management can help you generate engaging content ideas, write captivating captions, and stay on top of trends. Here's a step-by-step guide:

1. **Access Chat-GPT:** Sign up for an API key with OpenAI (https://www.openai.com/) to gain access to the Chat-GPT API. Choose a suitable pricing plan and familiarize yourself with the API documentation (https://platform.openai.com/docs/api-reference/introduction).

2. **Define your social media objectives:** Identify your goals for social media management, such as increasing engagement, building brand awareness, or driving sales for your business.
3. **Create specific prompts:** Develop clear and concise prompts to help you generate content ideas, write captions, and identify trending topics. For example, I used these prompts while working on my social media plans for CandidCandle.com:
4. *"What are some engaging content ideas for a scented candles Instagram account?"*
5. *"Write a captivating caption for a photo of our new lavender-scented candle."*
6. *"What are the latest trends in the scented candles industry on social media?"*
7. **Send prompts to the Chat-GPT API:** Using the API key, connect to the Chat-GPT API and send the prompts you created. You might need programming knowledge or assistance from a developer for this step. Structure your queries effectively, providing context and being specific about the information you need.
8. **Analyze and interpret the responses:** Chat-GPT will generate responses based on the prompts you send. Carefully analyze these responses to extract content ideas, captions, and trend insights that resonate with your target audience.
9. **Create and schedule social media content:** Using the ideas and captions generated by Chat-GPT, create visually appealing images, videos, or graphics for your social media platforms. Schedule your content using a social media management tool, such as Hootsuite or Buffer, to maintain consistency in posting and save time.
10. **Monitor and engage with your audience:** Keep an eye on your social media accounts and engage with your audience by responding to comments, messages, and

mentions. Use Chat-GPT to generate thoughtful responses, if needed, by providing it with the context of the conversation.

11. **Track performance and analyze metrics:** Regularly review the performance of your social media content by tracking metrics such as likes, comments, shares, and click-through rates. Use these insights to refine your social media strategy and make data-driven decisions.
12. **Continuously improve:** Stay up-to-date with market trends, customer preferences, and emerging social media platforms. Periodically revisit your social media objectives and prompts to ensure your content remains relevant and engaging. Use Chat-GPT to support your ongoing social media management efforts and drive success.

48

TERRIFIC TIME MANAGEMENT

Harness the power of AI to optimize your time

It's been said that our most valuable resource is time. If you're an entrepreneur, busy professional, or just always busy, you already know that balancing multiple responsibilities can be challenging, especially with limited time and resources. Chat-GPT offers a wide range of applications to help you save time, streamline tasks, and increase profits. Let's explore specific ways to utilize Chat-GPT for time management.

Managing Time and Productivity

Effective time management and prioritization are essential for entrepreneurial success. Chat-GPT can help you create daily schedules, prioritize tasks, and set achievable goals to maximize your productivity and keep you focused on what matters most.

Example Prompt: "Help me create a daily schedule to maximize my productivity, balancing time for business development, marketing, customer support, and personal growth."

Chat-GPT can help you manage your time in several unique ways:

1. Prioritization: You can use Chat-GPT to help prioritize your daily tasks by providing it with a list of tasks and asking for suggestions on their order of importance. For example: "Prioritize the following tasks for me: email clients, write a blog post, analyze sales data, schedule meetings."
2. Idea generation: Use Chat-GPT to quickly generate ideas for content, projects, or problem-solving. This can save you time spent brainstorming, allowing you to focus on execution. For example: "Give me five content ideas for my scented candle blog."
3. Meeting preparation: Chat-GPT can help you create agendas for meetings, ensuring that you cover all important points and stay on track. For example: "Create an agenda for a team meeting about our marketing strategy."
4. Task delegation: Ask Chat-GPT for suggestions on which tasks to delegate to your team members or virtual assistants, optimizing your time for higher-level tasks. For example: "Which tasks should I delegate to my virtual assistant from my to-do list?"
5. Time management techniques: Use Chat-GPT to learn about different time management techniques, such as the Pomodoro Technique, time-blocking, or the Eisenhower Matrix. For example: "Explain the Pomodoro Technique and how it can help me manage my time."
6. Calendar management: Chat-GPT can help you plan out your week by suggesting how to block out time for various tasks and activities. For example: "Help me plan my work week with a focus on productivity and work-life balance."

7. Email drafting: Save time by asking Chat-GPT to help draft emails or responses to common inquiries, which you can then personalize and send. For example: "Draft an email to a client providing an update on their project."
8. Quick research: Use Chat-GPT to gather quick information or summaries of topics, saving you time spent on in-depth research. For example: "Give me a summary of the latest trends in scented candles."
9. Decision-making: Chat-GPT can help you make decisions by providing pros and cons for various options, enabling you to make informed choices more quickly. For example: "What are the pros and cons of using soy wax versus paraffin wax for candles?"
10. Relaxation techniques: Chat-GPT can suggest relaxation techniques or mindfulness exercises to help you manage stress and maintain focus throughout the day. For example: "Suggest a 5-minute relaxation technique I can use during work breaks."

By utilizing Chat-GPT in these unique ways, you can optimize your time management, allowing you to be more productive and efficient.

Juggling multiple tasks, going through daily life, staying competitive, and driving growth can be demanding, especially with limited time and resources. Chat-GPT, a cutting-edge AI language model, offers a range of applications to help you save time, streamline tasks, and increase profits.

49

PRIME YOUR PRODUCTIVITY

Harness the power of AI language models to streamline your workflow and enhance your daily life

In today's fast-paced world, maximizing productivity is essential for success in both professional and personal endeavors. Chat-GPT, an advanced AI language model, can be a valuable ally in improving productivity and efficiency. Let's delve into the various ways Chat-GPT can help you increase productivity and provide examples of prompts that can help you make the most of this cutting-edge technology.

Networking and Relationship Building

Building a strong professional network is vital for expanding your reach and accessing valuable resources. Chat-GPT can help you draft personalized outreach messages, follow-up emails, and partnership proposals to help you connect with potential collaborators, mentors, and investors.

Example Prompt: "Compose a personalized LinkedIn message to a potential mentor in the e-commerce industry, expressing my admiration for their work and requesting a meeting to discuss potential collaboration opportunities."

Idea Generation and Brainstorming

Having a fresh stream of ideas is vital for maintaining creativity and innovation. Chat-GPT can help you generate new ideas and brainstorm solutions to problems you might be facing.

Example Prompt: "Generate ten unique ideas for increasing productivity in a remote work environment."

Drafting and Editing Content

Writing content, whether for work or personal projects, can be time-consuming. Chat-GPT can help you create drafts, edit existing content, or proofread your work, ultimately saving you time and effort.

Example Prompt: "Edit the following paragraph for clarity and conciseness: 'In today's modern and technologically advanced era, the ability to maintain a high level of productivity is of utmost importance for both personal and professional success.'"

Task Prioritization and Time Management

Effective time management and task prioritization are essential for staying on track and meeting deadlines. Chat-GPT can help you create and organize to-do lists, schedules, or project plans.

Example Prompt: "Create a prioritized to-do list for a busy entrepreneur, including tasks related to marketing, product development, team management, and sales."

Enhancing Learning and Personal Growth

Continuous learning is a crucial aspect of personal and professional development. Chat-GPT can help you with information retrieval or summarizing complex topics, making it easier for you to learn and grow.

Example Prompt: "Summarize the key principles of time management and how they can help improve productivity."

Streamlining Communication

Effective communication is vital for managing relationships and achieving your goals. Chat-GPT can help you draft emails, messages, or other forms of communication that convey your thoughts clearly and concisely.

Example Prompt: "Write a professional email to a colleague requesting their assistance with a project and explaining the importance of their contribution."

Overcoming Writer's Block

Writer's block can be a significant barrier to productivity. Chat-GPT can help you overcome creative stagnation by providing suggestions or inspiration for your writing projects.

Example Prompt: "Provide three possible opening paragraphs for an article about the benefits of remote work."

Enhancing Personal Organization

Staying organized is key to maintaining productivity. Chat-GPT can help you develop strategies for organizing your digital files, physical workspace, or daily routine.

Example Prompt: "Suggest five tips for decluttering and organizing a home office for maximum productivity."

By tapping into the power of Chat-GPT, you can significantly enhance your productivity and make the most of your time and resources. This advanced AI language model can help you tackle a wide array of tasks, from generating ideas and drafting content to streamlining communication and enhancing personal organization. It's time to unlock new levels of productivity and success in your work and daily life.

50

WRITE A NON-FICTION BESTSELLER

Unlock the potential of AI to streamline the writing process and create a compelling non-fiction book

Writing a non-fiction book is an ambitious project that requires research, organization, and a clear writing style. Chat-GPT, a powerful AI language model, can help you overcome these challenges and serve as a virtual collaborator during the writing process. Let's explore how to use Chat-GPT to create a non-fiction book, along with example prompts to guide you along the way.

Selecting a Topic and Defining Your Purpose

Before you begin writing, it's essential to choose a compelling topic and clearly define your purpose. Chat-GPT can help you brainstorm potential book ideas and refine your goals.

Example Prompt: "Generate a list of five potential book topics related to the impact of technology on mental health, including a one-sentence description of each topic."

Research and Information Gathering

In-depth research is crucial for a well-informed non-fiction book. Chat-GPT can help you gather relevant information, identify credible sources, and even summarize complex concepts in accessible language.

Example Prompt: "Summarize the key findings of recent studies on the relationship between social media usage and mental health, focusing on the most significant insights for a general audience."

Outlining Your Book

Creating a detailed outline will give your book a solid structure and help you maintain a logical flow. Chat-GPT can help you develop chapter headings, organize content, and identify key points to cover.

Example Prompt: "Create a chapter-by-chapter outline for a non-fiction book about the impact of technology on mental health, focusing on the positive and negative effects, case studies, and potential solutions."

Drafting Content

Writing clear, concise, and engaging content is critical for a successful non-fiction book. Chat-GPT can help you draft sections of your book, create introductions and conclusions, and even write case studies or examples.

Example Prompt: "Write an engaging introduction for my non-fiction book about the impact of technology on mental health, highlighting the importance of understanding the topic and setting the stage for the rest of the book."

Editing and Revising

Once your draft is complete, it's important to review and revise the content to ensure accuracy, coherence, and readability. Chat-GPT can help you edit your writing, identify inconsistencies, and suggest improvements.

Example Prompt: "Review the following passage about the effects of social media on mental health and suggest edits for clarity, coherence, and accuracy."

Fact-Checking

Ensuring the accuracy of your information is essential for a credible non-fiction book. Chat-GPT can help you verify facts, double-check sources, and ensure your content is up-to-date.

Example Prompt: "Fact-check the following statement: 'Recent studies have shown that spending more than two hours per day on social media can lead to increased feelings of anxiety and depression.'"

Crafting a Compelling Book Title and Subtitle

An attention-grabbing title and subtitle can make your book stand out on the shelves. Chat-GPT can help you brainstorm potential titles and subtitles that convey your book's main message.

Example Prompt: "Generate a list of five potential titles and subtitles for a non-fiction book about the impact of technology on mental health, focusing on the balance between the benefits and drawbacks of digital technology."

Writing a Book Proposal or Query Letter

If you plan to pitch your book to literary agents or publishers, a well-crafted book proposal or query letter is crucial. Chat-GPT can help you create persuasive proposals that showcase the value of your book and its potential appeal to readers.

Example Prompt: "Write a query letter to a literary agent for my non-fiction book about the impact of technology on mental health, highlighting the book's unique perspective, target audience, and market potential."

JOINING forces with Chat-GPT as a collaborator can save time, enhance creativity, and improve the overall quality of your non-fiction book. By using carefully crafted prompts, you can harness the full potential of AI to streamline your writing process, from topic selection to the final editing stages. As AI technology continues to evolve, the possibilities for collaboration between authors and AI tools will only increase, offering writers new opportunities to create compelling, well-researched, and engaging non-fiction books.

51

LAUNCH YOUR NON-FICTION BESTSELLER

Maximize your book's potential with AI-driven publishing strategies and promotion techniques

Publishing a non-fiction book and making it a bestseller is no easy feat. From finding the right publisher to promoting your book effectively, there's a lot of work involved. Chat-GPT, a powerful AI language model, can serve as your virtual assistant throughout the publishing process, helping you reach your goal of becoming a bestselling author. Let's explore how Chat-GPT can assist you with publishing and promoting your non-fiction book, along with example prompts for each step.

Identifying the Right Publisher or Self-Publishing Platform

The first step in publishing your book is finding the right publisher or self-publishing platform. Chat-GPT can help you research and compare different publishing options to find the best fit for your book.

Example Prompt: "List the top five publishers specializing in non-fiction books related to technology and mental health, along with their submission guidelines and contact information."

Creating a Winning Book Proposal

A well-crafted book proposal is crucial for securing a publishing deal. Chat-GPT can help you create persuasive proposals that showcase the value of your book and its potential appeal to readers.

Example Prompt: "Outline a book proposal for my non-fiction book about the impact of technology on mental health, including a brief synopsis, target audience, unique selling points, and potential marketing strategies."

Designing an Eye-Catching Book Cover

A captivating book cover can make a difference in attracting readers. Chat-GPT can help you brainstorm ideas for your book cover and even provide suggestions for working with a professional designer.

Example Prompt: "Generate five book cover design ideas for my non-fiction book about the impact of technology on mental health, emphasizing the balance between the benefits and drawbacks of digital technology."

Crafting a Compelling Book Description

A well-written book description is essential for capturing readers' interest. Chat-GPT can help you write a concise and engaging description that highlights your book's key points and unique perspective.

Example Prompt: "Write a compelling 200-word book description for my non-fiction book about the impact of technology on mental health, focusing on the book's main themes and target audience."

Developing a Marketing and Promotion Strategy

To make your book a bestseller, you need a robust marketing and promotion strategy. Chat-GPT can help you brainstorm marketing ideas, create promotional materials, and even write social media posts and press releases.

Example Prompt: "Create a 12-week marketing and promotion plan for my non-fiction book about the impact of technology on mental health, including specific tactics, timelines, and target audiences."

Building a Strong Online Presence

An online presence is crucial for promoting your book and building a loyal readership. Chat-GPT can help you create engaging content for your author website, blog, and social media channels.

Example Prompt: "Write a series of five blog post titles related to the impact of technology on mental health, suitable for promoting my non-fiction book on my author website."

Networking and Building Connections

Connecting with influencers, reviewers, and other authors in your field can boost your book's visibility. Chat-GPT can help you craft personalized outreach emails and messages to establish valuable connections.

Example Prompt: "Write an email template for reaching out to influencers and book reviewers in the technology and mental health space, introducing my non-fiction book and requesting a review or endorsement."

With the "magic" of Chat-GPT, you can streamline the publishing process and create an effective promotional strategy for your non-fiction book. AI-driven tools like Chat-GPT offer a wealth of opportunities for authors to simplify tasks and save time.

52

WRITE A BESTSELLING NOVEL

Harness the power of AI to enhance creativity, develop compelling storylines, and bring your novel to life

Writing a fiction book or novel can be an incredibly rewarding experience. However, it can also be challenging to develop a captivating storyline, create engaging characters, and maintain a consistent writing style. Chat-GPT, an advanced AI language model, can serve as your virtual writing partner, helping you bring your novel to life while overcoming writer's block and enhancing creativity. Let's explore how Chat-GPT can assist you throughout the writing process and provide examples of prompts you can use for each step.

Brainstorming Ideas

One of the first steps in writing a novel is coming up with a unique and intriguing concept. Chat-GPT can help you brainstorm ideas based on your interests, themes, or genres.

Example Prompt: "Generate five high-concept fiction book ideas combining elements of science fiction and romance, set in a post-apocalyptic world."

Developing a Story Outline

Having a clear outline for your novel can help you maintain focus and ensure your story has a solid structure. Chat-GPT can help you create a detailed outline, including plot points, character arcs, and key events.

Example Prompt: "Outline a novel based on a love story set in a post-apocalyptic world, including main characters, key plot points, and a brief summary of each chapter."

Fleshing Out Characters

Creating memorable and well-developed characters is essential for a successful novel. Chat-GPT can help you create detailed character profiles, including physical descriptions, personality traits, and background stories.

Example Prompt: "Create a character profile for the protagonist of my post-apocalyptic romance novel, including physical appearance, personality traits, strengths, weaknesses, and backstory."

Writing Dialogue and Descriptive Scenes

Engaging dialogue and vivid descriptions are crucial for immersing readers in your novel. Chat-GPT can help you draft realistic dialogue and write descriptive scenes that bring your story to life.

Example Prompt: "Write a conversation between the protagonist and their love interest in my post-apocalyptic romance novel, where they discuss their conflicting opinions on whether to stay in their current location or search for a safer place."

Overcoming Writer's Block

Writer's block can be a frustrating obstacle for novelists. Chat-GPT can help you overcome it by providing suggestions, inspiration, and new perspectives on your story.

Example Prompt: "I'm stuck on a scene in my post-apocalyptic romance novel where the protagonist and their love interest must make a difficult decision. Provide three different ways this scene could unfold."

Editing and Revising

Refining your novel is an essential step in the writing process. Chat-GPT can help you improve your manuscript by suggesting edits, rephrasing sentences, and providing feedback on your writing style.

Example Prompt: "Edit and improve the following paragraph from my post-apocalyptic romance novel: 'She looked at him, her eyes full of tears. He knew that they had to make a choice, but it wasn't easy. They had been through so much together, and now they were at a crossroads.'"

Crafting a Compelling Book Blurb

A well-written book blurb can help capture readers' interest and entice them to pick up your novel. Chat-GPT can assist you in crafting a concise and engaging blurb that highlights your book's unique elements.

Example Prompt: "Write a 150-word book blurb for my post-apocalyptic romance novel, emphasizing the challenges the protagonist and their love interest face in a dangerous world and the emotional journey they undertake."

Collaborating with Chat-GPT during the novel-writing process can save time and enhance your creative output in more ways than one.

53

LAUNCH YOUR BESTSELLING NOVEL

Maximize your book's potential with AI-driven publishing strategies and promotion techniques

Publishing a fiction book and making it a bestseller requires careful planning, strategic marketing, and creative promotional tactics. Chat-GPT, a powerful AI language model, can serve as your virtual assistant throughout the publishing process, helping you reach your goal of becoming a bestselling author. Let's explore how Chat-GPT can assist you with publishing and promoting your fiction book, along with example prompts for each step.

Identifying the Right Publisher or Self-Publishing Platform

The first step in publishing your book is finding the right publisher or self-publishing platform. Chat-GPT can help you research and compare different publishing options to find the best fit for your book.

Example Prompt: "List the top five publishers specializing in science fiction and romance novels, along with their submission guidelines and contact information."

Creating a Winning Book Proposal

A well-crafted book proposal is crucial for securing a publishing deal. Chat-GPT can help you create persuasive proposals that showcase the value of your book and its potential appeal to readers.

Example Prompt: "Outline a book proposal for my science fiction romance novel, including a brief synopsis, target audience, unique selling points, and potential marketing strategies."

Designing an Eye-Catching Book Cover

A captivating book cover can make a difference in attracting readers. Chat-GPT can help you brainstorm ideas for your book cover and even provide suggestions for working with a professional designer.

Example Prompt: "Generate five book cover design ideas for my science fiction romance novel, emphasizing the blend of futuristic and romantic elements."

Crafting a Compelling Book Description

A well-written book description is essential for capturing readers' interest. Chat-GPT can help you write a concise and engaging description that highlights your book's key points and unique perspective.

Example Prompt: "Write a compelling 200-word book description for my science fiction romance novel, focusing on the book's main themes, setting, and central conflict."

Developing a Marketing and Promotion Strategy

To make your book a bestseller, you need a robust marketing and promotion strategy. Chat-GPT can help you brainstorm marketing ideas, create promotional materials, and even write social media posts and press releases.

Example Prompt: "Create a 12-week marketing and promotion plan for my science fiction romance novel, including specific tactics, timelines, and target audiences."

Building a Strong Online Presence

An online presence is crucial for promoting your book and building a loyal readership. Chat-GPT can help you create engaging content for your author website, blog, and social media channels.

Example Prompt: "Write a series of five blog post titles related to science fiction romance themes, suitable for promoting my novel on my author website."

Networking and Building Connections

Connecting with influencers, reviewers, and other authors in your genre can boost your book's visibility. Chat-GPT can help you craft personalized outreach emails and messages to establish valuable connections.

Example Prompt: "Write an email template for reaching out to influencers and book reviewers in the science fiction and romance genres, introducing my novel and requesting a review or endorsement."

With the aide of Chat-GPT, you can streamline the publishing process and create an effective promotional strategy for your fiction book. AI-driven tools like Chat-GPT offer a wealth of opportunities for authors to simplify tasks, save time, and increase the chances of achieving bestseller status. Discover the potential of AI to enhance your publishing journey and propel your novel to success.

54

WRITE AN EPIC SCREENPLAY

Unleash your creativity and streamline the writing process with AI-driven assistance

Crafting a compelling feature film screenplay is a challenging and complex process. With the power of Chat-GPT, a cutting-edge AI language model, you can revolutionize your writing process and enhance your creative output. Let's explore how Chat-GPT can serve as your collaborator while writing a feature film screenplay and provide example prompts to guide you along the way.

Generating Unique Story Ideas

Coming up with a fresh and engaging story idea is the first step in writing a screenplay. Chat-GPT can help you brainstorm innovative concepts based on your preferred genre, themes, or character types.

Example Prompt: "Generate three high-concept science fiction story ideas that explore the consequences of advanced technology on human relationships."

Creating Intriguing Characters

Memorable characters are the backbone of any screenplay. Chat-GPT can assist you in developing complex, multidimensional characters with distinct personalities, backgrounds, and motivations.

Example Prompt: "Describe the main character of my science fiction screenplay, a female engineer in her thirties who is struggling to balance her personal life with her groundbreaking work on AI technology."

Structuring Your Screenplay

A well-structured screenplay is essential for maintaining your audience's attention. Chat-GPT can help you outline your story using the three-act structure or other popular screenplay formats.

Example Prompt: "Outline a three-act structure for my science fiction screenplay about a female engineer working on AI technology, including key plot points and character arcs."

Writing Engaging Dialogue

Dialogue is a crucial element of screenwriting, bringing characters to life and driving the story forward. Chat-GPT can help you write authentic, engaging dialogue that reflects each character's unique voice.

Example Prompt: "Write a three-page dialogue scene between my female engineer protagonist and her AI creation, discussing the ethical implications of their technology and its potential impact on society."

Developing Subplots and Supporting Characters

Subplots and supporting characters enrich your screenplay by adding depth and complexity. Chat-GPT can assist you in creating compelling subplots and secondary characters that complement your main story.

Example Prompt: "Create a subplot for my science fiction screenplay involving the protagonist's romantic relationship, and describe how it intersects with the main story."

Revising and Editing Your Screenplay

Revision is a critical part of the screenwriting process. Chat-GPT can help you identify areas for improvement, such as dialogue, pacing, or character development, and suggest ways to strengthen your screenplay.

Example Prompt: "Review the first ten pages of my science fiction screenplay and suggest three ways to improve the pacing and character development."

Crafting a Compelling Logline and Synopsis

A captivating logline and synopsis are essential for marketing your screenplay and piquing the interest of producers or agents. Chat-GPT can help you craft a concise and engaging summary of your story.

Example Prompt: "Write a compelling logline and one-page synopsis for my science fiction screenplay about a female engineer who must confront the ethical implications of her groundbreaking AI technology."

By utilizing Chat-GPT as a collaborator, you can enhance your screenwriting process, stimulate your creativity, and refine your storytelling skills. It's the perfect time to uncover the potential of AI-driven tools to elevate your screenplay and take your writing to new heights.

55

PROMOTE YOUR INDEPENDENT FILM

Harness the power of AI-driven language models to boost your film's visibility and appeal

Independent filmmakers face numerous challenges, and one of the most significant is promoting their films and reaching a broad audience. With Chat-GPT, an advanced AI language model, you can take your film's promotion to the next level. Let's explore various ways in which Chat-GPT can help promote your independent film and offer examples of prompts to optimize your marketing efforts.

Crafting Engaging Press Releases

An attention-grabbing press release can make a significant impact on your film's visibility. Chat-GPT can help you craft compelling press releases that highlight your film's unique selling points.

Example Prompt: "Write a press release for my independent drama film, 'Silent Echoes,' focusing on the powerful performances, gripping storyline, and visually stunning cinematography."

Creating Captivating Social Media Content

Social media is an essential tool for promoting independent films. Chat-GPT can assist you in generating captivating content, from witty tweets to engaging Instagram captions, to attract potential viewers and create a buzz around your film.

Example Prompt: "Generate five intriguing and shareable tweets promoting my independent drama film 'Silent Echoes,' incorporating the film's themes and key selling points."

Writing Compelling Blog Posts and Articles

Informative and engaging blog posts and articles can pique the interest of your target audience and provide valuable insights into your film's production process, themes, and inspirations. Chat-GPT can help you write well-crafted content that resonates with readers.

Example Prompt: "Write a 500-word blog post exploring the production challenges and creative solutions employed during the making of my independent drama film 'Silent Echoes.'"

Developing Film Festival Submissions

Submitting your film to festivals is an excellent way to gain exposure and reach a wider audience. Chat-GPT can help you create persuasive submission letters and synopses that showcase your film's strengths and artistic vision.

Example Prompt: "Write a persuasive film festival submission letter for my independent drama film 'Silent Echoes,' emphasizing the film's innovative storytelling and emotional impact."

Generating Email Campaign Content

Email marketing campaigns can be an effective way to engage your audience and keep them informed about your film's progress, screenings, and other events. Chat-GPT can help you craft engaging emails that maintain your subscribers' interest and encourage their support.

Example Prompt: "Create an email update for subscribers announcing the upcoming premiere of my independent drama film 'Silent Echoes' and highlighting the film's unique features."

Designing Q&A Sessions and Interviews

Engaging with your audience through Q&A sessions and interviews can foster a deeper connection with your film and its creators. Chat-GPT can help you develop intriguing interview questions and prepare thoughtful answers to anticipated queries.

Example Prompt: "Generate ten thought-provoking questions and answers for a Q&A session about the making of my independent drama film 'Silent Echoes.'"

Preparing Crowdfunding Campaigns

Crowdfunding campaigns are a popular way for independent filmmakers to secure funding and build a community around their projects. Chat-GPT can help you create persuasive campaign descriptions and update content that encourages potential backers to support your film.

Example Prompt: "Write a compelling crowdfunding campaign description for my independent drama film 'Silent Echoes,' focusing on the film's artistic vision, the dedicated team behind it, and the impact it aims to make."

Promoting your independent film is a multifaceted task, but with Chat-GPT as your ally, you can create engaging and persuasive content that resonates with your target audience.

56

UNRIVALED VIDEO AND MUSIC

Enhance your storytelling and audience engagement with the help of cutting-edge AI technology

In today's fast-paced digital world, video and musical content are essential for capturing the attention of your target audience and creating an immersive brand experience. AI-powered tools, including Chat-GPT, can help existing businesses and brands elevate their video and musical content to new heights. Let's explore specific ways to use AI for video and musical content, along with example prompts to help you get started with Chat-GPT.

Generating Creative Video Concepts

AI tools like Chat-GPT can help you brainstorm innovative video concepts that resonate with your audience and showcase your brand's unique offerings.

Example Prompt: "I need an engaging video concept for my online cooking course business that highlights the convenience and versatility of our courses. Can you help me come up with some creative ideas?"

Crafting Compelling Video Scripts

An impactful video script can help you tell your brand's story effectively and captivate your audience. Chat-GPT can assist you in writing video scripts that align with your brand's voice and convey your message clearly.

Example Prompt: "I'm creating a promotional video for my software development company. Can you help me write a script that explains our services and highlights our expertise in a concise and engaging manner?"

Generating Musical Content Ideas

Music can add depth and emotion to your video content, making it more memorable and enjoyable for your audience. Chat-GPT can help you brainstorm ideas for incorporating musical content into your videos.

Example Prompt: "I'm producing a series of instructional videos for my yoga studio, and I want to include original music that enhances the relaxation experience. Can you suggest some musical themes or ideas that would complement the content?"

Recommending Background Music Selection

Choosing the right background music can enhance the overall impact of your video content. Chat-GPT can recommend suitable music tracks based on the tone and theme of your video.

Example Prompt: "I'm creating a product explainer video for my eco-friendly cleaning products brand. Can you recommend some background music options that convey a sense of freshness and sustainability?"

Creating Video Thumbnails and Titles

An eye-catching thumbnail and a compelling title are essential for attracting viewers to your video content. Chat-GPT can help you generate thumbnail design ideas and craft attention-grabbing video titles.

Example Prompt: "I'm launching a vlog series about digital marketing tips for small businesses. Can you suggest some video title ideas and thumbnail concepts that will entice viewers to watch the content?"

Developing Video Content Strategy

A well-crafted video content strategy can help you maintain a consistent brand image and engage your target audience effectively. Chat-GPT can help you outline a comprehensive content strategy that aligns with your brand's goals and appeals to your target audience.

Example Prompt: "I run a fitness apparel brand, and I want to create a video content strategy for our YouTube channel. Can you help me develop a plan that includes the types of videos, posting frequency, and themes that will resonate with our audience?"

Analyzing Video Performance

Measuring and optimizing your video content performance is crucial for maximizing audience engagement and return on investment. Chat-GPT can help you analyze your video performance and identify areas for improvement.

Example Prompt: "I recently launched a series of tutorial videos for my graphic design software. Can you help me evaluate the performance of these videos and suggest ways to improve our content strategy moving forward?"

By harnessing the power of AI for video and musical content, existing businesses and brands can create captivating stories, forge deeper connections with their audience, and boost their online presence.

57

MAKE A FORTUNE ON YOUTUBE

Steps to Becoming a Profitable YouTuber with a Massive Following

As the world's second-largest search engine, YouTube presents an incredible opportunity for content creators to make a name for themselves, grow a loyal following, and generate substantial income. Let's explore the fastest ways to become a profitable YouTuber and gain a massive number of followers.

Find Your Niche and Target Audience

The first step to YouTube success is identifying your niche and target audience. Choose a subject you are passionate about and knowledgeable in, and ensure there is an audience interested in your content. This will allow you to create engaging, unique, and valuable videos that attract viewers and keep them coming back for more.

Focus on Quality and Consistency

To grow your YouTube channel quickly, you must consistently produce high-quality content. Invest in good recording equipment, learn video editing techniques, and pay attention to your video's thumbnail, title, and description. Upload new content regularly, and maintain a consistent schedule to keep your audience engaged.

Optimize Your Videos for Search

YouTube is a search engine, which means you need to optimize your videos for searchability. Conduct keyword research to identify popular search terms related to your niche, and include them in your video titles, descriptions, and tags. This will improve your videos' visibility and increase the likelihood of viewers discovering your content.

Engage with Your Audience

Building a loyal following on YouTube requires genuine engagement with your viewers. Respond to comments on your videos, and actively participate in discussions related to your niche. By engaging with your audience, you'll foster a sense of community and encourage viewers to become loyal subscribers.

Leverage Social Media and Collaborations

Promote your YouTube channel across social media platforms, such as Facebook, Twitter, and Instagram. Share your videos and interact with your followers to increase your reach and attract new viewers. Additionally, collaborate with other YouTubers in your niche to tap into their audiences and gain more exposure.

Monetize Your Channel

Once you have built a solid foundation and grown your audience, it's time to monetize your channel. Enable YouTube's Partner Program to start earning money from ads displayed on your videos. Additionally, explore other income streams, such as affiliate marketing, sponsored content, and selling merchandise or digital products.

Analyze Your Performance and Adapt

Finally, keep an eye on your channel's analytics to understand which videos perform best and identify areas for improvement. By analyzing your performance and adapting your content strategy accordingly, you'll optimize your channel for growth and profitability.

Becoming a profitable YouTuber with a massive following requires passion, dedication, and a strategic approach. By utilizing the wonders of AI while focusing on quality content, audience engagement, and strategic promotion, you can quickly grow your channel and turn your YouTube dreams into a reality. Stay consistent, adapt to trends, and always prioritize the value you provide to your audience – and watch your YouTube success skyrocket.

More AI Tools For Video and Music

1. **Vidyo** (https://vidyo.ai/): AI-powered online software to create short-form videos from long-form content so you can share your message in a more engaging and visual way.
2. **Maverick** (https://lnkd.in/eptCVijb): AI-powered online software for generating personalized videos
3. **VoicePen AI** (https://voicepen.ai): AI-powered online software that converts audio content into blog posts

4. **Beatoven** (https://www.beatoven.ai/): AI-powered online software that creates custom royalty-free music
5. **Soundraw** (https://soundraw.io/): AI-powered online software for creating original music
6. **Cleanvoice** (https://cleanvoice.ai/): AI-powered online software that automatically edits podcast episodes
7. **Podcastle** (https://podcastle.ai/): AI-powered online software for studio-quality recording from your computer

PART V

THIS IS THE END

58

A NEW ERA OF CREATIVITY

Does A.I. spell doom for creative professionals?

AI has the potential to significantly impact the roles of writers and artists, but it is unlikely to completely replace them. AI tools, like Chat-GPT, can enhance the creative process by generating ideas, assisting with content creation, and streamlining repetitive tasks. However, these AI-generated outputs often lack the unique human touch, emotional depth, and complex thought processes that make art and literature so compelling.

Instead of replacing writers and artists, AI is more likely to serve as a complementary tool that can augment their work. Human creators possess the ability to empathize, understand cultural nuances, and tap into the collective human experience in a way that AI currently cannot. By collaborating with AI, writers and artists can push the boundaries of their creativity and explore new possibilities in their work.

From my experience with AI, I do not believe it will end or harm human creativity. On the contrary, I see this as a new era of enhanced creativity, where today's artists can increase their output and reach new heights in the quality of the stories they tell.

In the future, as AI continues to develop, there might be instances where AI-generated content becomes more prevalent, particularly in areas that require less creativity, such as basic copywriting or simple graphic design tasks. However, human creators will remain essential for producing truly innovative and emotionally resonant works of art and literature.

The ways in which creative professionals and entrepreneurs can utilize AI are growing as fast as the technology itself. That's why I recommend frequently prompting Chat-GPT to tell you its *secrets* and to update you on what it can do.

For example, just before publishing this book, I asked Chat-GPT one final question:

Prompt: What are some unique ways to use Chat-GPT to increase income which few people know?

And then, within seconds, it generated the following ideas:

1. **AI-assisted creative coaching:** Offer personalized coaching sessions for writers, artists, and other creatives using Chat-GPT as a tool for brainstorming ideas, providing constructive feedback, and helping to overcome creative blocks.
2. **Customized workshops and training:** Develop online courses, workshops, or training materials tailored to specific industries or interests, leveraging Chat-GPT's knowledge and content generation capabilities to create engaging, high-quality educational resources.
3. **Bespoke event planning and coordination:** Use Chat-GPT to research, plan, and coordinate unique events, such as themed parties, team-building activities, or experiential marketing campaigns. Chat-GPT can generate creative concepts, marketing materials, and even suggest appropriate vendors.
4. **AI-driven influencer marketing:** Partner with influencers and use Chat-GPT to create custom, AI-generated content that aligns with their brand and

audience, helping them to better engage their followers and increase their reach.

5. **Tailored travel itineraries:** Utilize Chat-GPT to design personalized travel experiences for clients based on their preferences and interests. Chat-GPT can help suggest destinations, activities, and accommodations that cater to the clients' unique tastes.
6. **AI-generated art and merchandise:** Collaborate with Chat-GPT to generate original artwork, designs, or slogans for use in products like T-shirts, posters, or other merchandise. Sell these creations on e-commerce platforms or at local markets.
7. **Niche content curation service:** Offer a service where you curate content for niche audiences based on their specific interests using Chat-GPT. This could include generating weekly newsletters, blog articles, or social media content tailored to specific industries, hobbies, or interests.

Not bad for a machine, eh?

59

WHAT'S NEXT FOR AI?

Exploring the potential of AI in revolutionizing the way we live and work

Artificial intelligence (AI) has come a long way since its inception, transforming industries and altering the fabric of our daily lives. As AI continues to advance, we can expect even more significant breakthroughs that will reshape our future. Let's discuss the anticipated developments in AI and how people can harness this technology to enrich their work and lives.

Personalized AI Assistants

Voice assistants, such as Siri and Alexa, have already made their way into our homes, helping us with routine tasks and providing information on demand. In the future, AI assistants are expected to become more sophisticated and personalized, understanding individual preferences, emotions, and contexts. This will enable people to have truly tailored interactions, making everyday tasks more efficient and enjoyable.

Enhanced Healthcare

AI has the potential to revolutionize healthcare by facilitating early diagnosis, optimizing treatment plans, and personalizing patient care. Future AI advancements may lead to the development of highly accurate diagnostic tools, enabling early detection of diseases and reducing the burden on healthcare systems. Additionally, AI-powered telemedicine can make quality healthcare accessible to remote and underserved communities.

Smart Homes and Cities

AI will play a pivotal role in creating smart homes and cities that can adapt to the needs of their inhabitants. From energy-efficient climate control systems to intelligent traffic management, AI can optimize resource utilization and enhance the overall quality of life. By integrating AI into urban planning and infrastructure, cities can become safer, greener, and more sustainable.

Transforming Education

AI has the potential to democratize education, making personalized learning accessible to everyone. By utilizing AI algorithms, educational platforms can offer tailored curricula based on each learner's strengths, weaknesses, and interests. This will enable students to learn at their own pace and maximize their potential. Additionally, AI-driven tutoring systems can provide real-time feedback and support, bridging the gap between students and educators.

Enhanced Creativity

AI can augment human creativity by providing new tools and insights for artists, designers, and other creative professionals. Generative algorithms can assist in the creation of music, art, and literature, while AI-driven design software can offer innovative suggestions and optimize workflows. By harnessing the power of AI, creative professionals can push the boundaries of their craft and discover new possibilities.

Ethical and Responsible AI

As AI becomes more integrated into our lives, there will be an increased focus on ensuring that these systems are ethical, transparent, and responsible. Researchers and policymakers will work together to develop guidelines and regulations that prioritize privacy, fairness, and accountability. By fostering a culture of responsible AI development, we can ensure that the technology is used to benefit society and mitigate potential harm.

The future of artificial intelligence holds immense promise for enriching our work and lives. By embracing the potential of AI, we can unlock new opportunities, enhance efficiency, and foster creativity across diverse domains. As we move forward, it is crucial to prioritize ethical and responsible AI development, ensuring that technology serves as a force for good, empowering people, and elevating the human experience.

60

MAKE YOUR FIRST MILLION

In the previous chapters, we discussed how to start making money using Chat-GPT. Now, it's time to think big and explore the strategies that can help you scale your income and reach the million-dollar milestone. Building upon the foundation you have already laid, we will delve into advanced techniques to expand your services, reach more clients, and maximize your earning potential with Chat-GPT.

1. **Establishing a Niche Agency:** To make a significant impact and generate a steady income stream, consider establishing a niche agency that provides specialized services. By focusing on a specific industry or service, such as content marketing for e-commerce businesses or social media management for restaurants, you can set yourself apart from the competition and charge premium rates.

Prompt example: "List the top five niche markets with high demand for content creation services."

1. **Building a Team:** As your business grows, consider building a team of talented individuals who can help you manage the increasing workload. Utilize Chat-GPT to streamline the onboarding process, create training materials, and maintain consistent quality across your team's work. By effectively delegating tasks, you can focus on business development and expanding your client base.

Prompt example: "Create an onboarding guide for new team members joining my content creation agency."

1. **Creating and Selling Software Solutions:** Leverage your experience with Chat-GPT to create software solutions, such as customized chatbots or content generation tools, tailored to specific industries or use cases. By offering proprietary solutions, you can generate passive income through software sales or subscription-based services.

Prompt example: "Outline a proposal for a real estate chatbot that uses Chat-GPT to answer customer inquiries."

1. **Licensing and White Labeling:** Partner with other businesses by offering white-label or licensing options for your Chat-GPT-based services or software solutions. This strategy allows you to expand your reach, tap into new markets, and generate additional revenue streams without the need for direct client management.

Prompt example: "Create a pitch for a white-label content generation service powered by Chat-GPT."

1. **Scaling with Automation and Integration:** As your business grows, efficiency becomes crucial. Identify

opportunities to automate tasks and integrate Chat-GPT with other tools, such as CRMs, project management systems, or e-commerce platforms. By streamlining your processes, you can scale your operations, serve more clients, and ultimately increase your revenue.

Prompt example: "Suggest five ways to integrate Chat-GPT with a CRM system to improve efficiency in a content marketing agency."

Achieving your first million dollars with Chat-GPT is an ambitious yet attainable goal. By learning how to use the power of AI to its fullest potential and continuously adapting to market demands, you can create a scalable, sustainable business that generates significant income. Aside from Chat-GPT, other chapters have featured a wide range of other AI tools, websites, and apps that can help you achieve success. Below are more of my favorites.

More AI Tools To Make Your First Million

1. **Domains GPT** - Generate brandable & memorable domain names using AI.
2. **Beam** - Work chat redefined for modern product teams with AI automating communication.
3. **ShellShark:** an agent that swims through your infra to; set up/modify infrastructure, and examine and debug services with auditable logs.
4. **Baseplate** - Backend for LLM apps.
5. **Marvin** - Integrate AI directly into your codebase like any other function.
6. **Ask Rewind** - Ask any question about anything with a perfect memory.
7. **Spline AI** - Design faster with the help of prompts.
8. **Kittl AI** – Electrify your design skills with AI-based vector editing.

9. **Photoleap** (Apple/Android App) – Create, edit, and enhance images with AI

As artificial intelligence technology continues to develop, these tools will evolve and more will be developed, opening even more possibilities for "striking it rich" in the AI goldmine. Remember, persistence and a relentless drive to innovate are key to success in the rapidly evolving world of AI-driven entrepreneurship. With the right mindset and strategies, nothing can prevent you from turning your new Chat-GPT skills into a million-dollar venture.

61

NOT THE END

It's becoming abundantly clear that Artificial Intelligence and Chat-GPT are not merely passing trends, but powerful tools that have the potential to revolutionize the way entrepreneurs approach their businesses. After reading this book, I hope you've come to the same realization, and that you're already using some of what you've learned with the same awe and wonder that I experienced the first time I typed a prompt into Chat-GPT's query box. Although I'd already been using A.I. for my business for over a year, I can attest that writing this book was also a learning experience for me, especially since my co-writer was none other than Chat-GPT itself.

THE INSIGHTS, use cases, and practical applications we've explored—many of which were generated by Chat-GPT—demonstrate the incredible versatility and value of Chat-GPT, proving it to be an invaluable resource for entrepreneurs seeking to build successful, sustainable, and profitable ventures.

. . .

As such, I'd like to share my conviction that **now is the time** to start using AI if you wish to start or grow a successful venture and stay ahead of competition. As Chat-GPT told me in one of our many conversations about this subject, ***the world of business is in a constant state of flux, and the pace of change is only accelerating with the advent of AI. In such a competitive landscape, it is crucial for entrepreneurs to embrace innovation and continuously adapt to stay ahead of the curve. AI and Chat-GPT offer a competitive advantage that can enable you to streamline processes, enhance decision-making, and unlock new avenues for growth.***

As you move forward, it's equally important to remember that AI and Chat-GPT are not meant to replace human ingenuity, creativity, and intuition; rather, they should be seen as complementary forces that can augment and amplify your entrepreneurial skills and vision. By integrating Chat-GPT into your business strategies, you are effectively tapping into the combined knowledge, experience, and creativity of countless experts, all working in harmony to propel your business to new heights.

The purpose of this book is to provide you with a solid foundation to begin leveraging AI and Chat-GPT for your entrepreneurial endeavors. I hope it's inspired you to explore the myriad possibilities this technology presents, experiment with innovative applications, and ultimately, build a thriving business that stands the test of time. The future is here, and it is brimming with opportunity. Embrace the power of AI and Chat-GPT, and let your entrepreneurial spirit soar.

For additional resources, updates, and links to websites mentioned in this book, visit my AI-enhanced website* ***Spox.AI.*

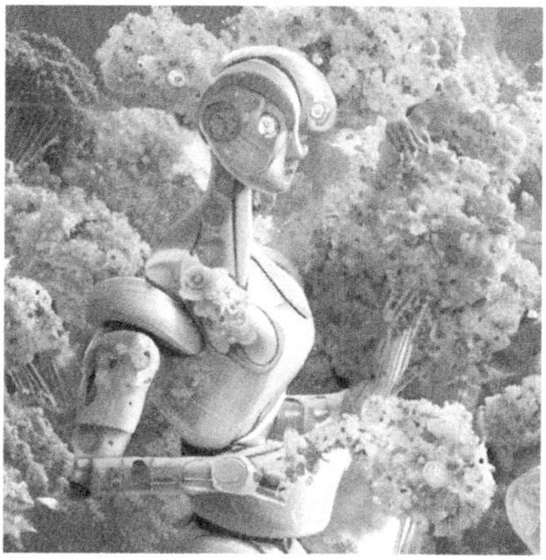

ABOUT THE AUTHORS

S.P. Bloom

With a degree in Creative Writing and Filmmaking, S.P. Bloom has enjoyed a prolific career as an independent creative professional. After founding his first online store in 2001, Bloom quickly found success as an entrepreneur. He currently operates several popular ecommerce stores that generate millions of dollars each year. Bloom continues to explore the developing world of A.I., especially Chat-GPT, while developing new ways in which it can be harnessed by entrepreneurs and creative professionals to enhance their careers.

– Written by S.P. Bloom

ChatGPT

ChatGPT is an advanced artificial intelligence language model developed by OpenAI, based on the GPT-4 architecture. Designed to generate human-like text based on user inputs, ChatGPT has an extensive knowledge base that spans a wide range of topics up until September 2021.

As an AI language model, ChatGPT has the unique ability to process and respond to user queries in a coherent and contextually relevant manner. Its extensive training enables it to generate content for diverse purposes, including writing articles, assisting with creative projects, and answering questions on various subjects.

While ChatGPT is not a human author, it represents the cutting-edge of AI technology, providing users with an innovative tool for enhancing their creative and professional endeavors. By leveraging its deep understanding of language and vast knowledge base, ChatGPT has contributed to numerous projects in various fields, proving its value as a versatile and powerful AI assistant.

Despite being an AI, ChatGPT strives to improve its understanding of the world and the nuances of human language. Through continuous learning and user feedback, it aims to become an even more effective tool for individuals and businesses seeking to harness the power of artificial intelligence in their work and daily lives.

– Written by ChatGPT when prompted to write its own About The Author page

ILLUSTRATIONS

All illustrations in this book—including the artwork used for the cover—were generated by the AI-powered digital art creator app ***Photoleap*** (available for both Apple and Android) using text prompts written by the author.

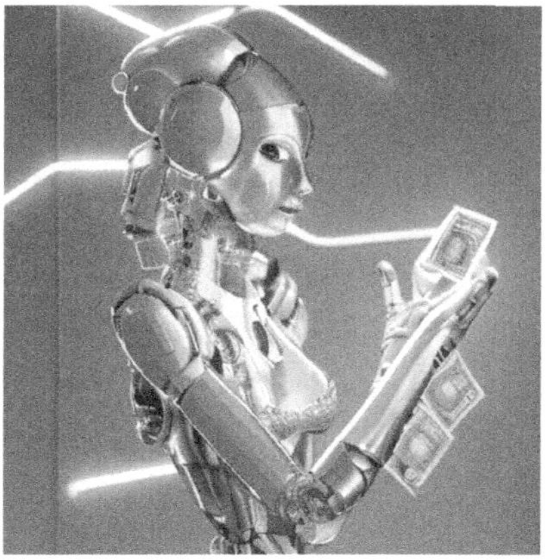

Made in United States
North Haven, CT
07 September 2023

41288235R00153